Pitman Research Notes in Mathematics Series

Main Editors
H. Brezis, Université de Paris
R. G. Douglas, State University of New York at Stony Brook
A. Jeffrey, University of Newcastle-upon-Tyne *(Founding Editor)*

Editorial Board
R. Aris, University of Minnesota
A. Bensoussan, INRIA, France
S. Bloch, University of Chicago
B. Bollobás, University of Cambridge
W. Bürger, Universität Karlsruhe
S. Donaldson, University of Oxford
J. Douglas Jr, University of Chicago
R. J. Elliott, University of Alberta
G. Fichera, Università di Roma
R. P. Gilbert, University of Delaware
R. Glowinski, Université de Paris
K. P. Hadeler, Universität Tübingen
K. Kirchgässner, Universität Stuttgart
B. Lawson, State University of New York at Stony Brook
W. F. Lucas, Claremont Graduate School
R. E. Meyer, University of Wisconsin-Madison
L. E. Payne, Cornell University
G. F. Roach, University of Strathclyde
J. H. Seinfeld, California Institute of Technology
B. Simon, California Institute of Technology
I. N. Stewart, University of Warwick
S. J. Taylor, University of Virginia

Submission of proposals for consideration
Suggestions for publication, in the form of outlines and representative samples, are invited by the Editorial Board for assessment. Intending authors should approach one of the main editors or another member of the Editorial Board, citing the relevant AMS subject classifications. Alternatively, outlines may be sent directly to the publisher's offices. Refereeing is by members of the board and other mathematical authorities in the topic concerned, throughout the world.

Preparation of accepted manuscripts
On acceptance of a proposal, the publisher will supply full instructions for the preparation of manuscripts in a form suitable for direct photo-lithographic reproduction. Specially printed grid sheets are provided and a contribution is offered by the publisher towards the cost of typing. Word processor output, subject to the publisher's approval, is also acceptable.

Illustrations should be prepared by the authors, ready for direct reproduction without further improvement. The use of hand-drawn symbols should be avoided wherever possible, in order to maintain maximum clarity of the text.

The publisher will be pleased to give any guidance necessary during the preparation of a typescript, and will be happy to answer any queries.

Important note
In order to avoid later retyping, intending authors are strongly urged not to begin final preparation of a typescript before receiving the publisher's guidelines and special paper. In this way it is hoped to preserve the uniform appearance of the series.

Longman Scientific & Technical
Longman House
Burnt Mill
Harlow, Essex, UK
(tel (0279) 26721)

Titles in this series

1. Improperly posed boundary value problems
   A Carasso and A P Stone
2. Lie algebras generated by finite dimensional ideals
   I N Stewart
3. Bifurcation problems in nonlinear elasticity
   R W Dickey
4. Partial differential equations in the complex domain
   D L Colton
5. Quasilinear hyperbolic systems and waves
   A Jeffrey
6. Solution of boundary value problems by the method of integral operators
   D L Colton
7. Taylor expansions and catastrophes
   T Poston and I N Stewart
8. Function theoretic methods in differential equations
   R P Gilbert and R J Weinacht
9. Differential topology with a view to applications
   D R J Chillingworth
10. Characteristic classes of foliations
    H V Pittie
11. Stochastic integration and generalized martingales
    A U Kussmaul
12. Zeta-functions: An introduction to algebraic geometry
    A D Thomas
13. Explicit *a priori* inequalities with applications to boundary value problems
    V G Sigillito
14. Nonlinear diffusion
    W E Fitzgibbon III and H F Walker
15. Unsolved problems concerning lattice points
    J Hammer
16. Edge-colourings of graphs
    S Fiorini and R J Wilson
17. Nonlinear analysis and mechanics: Heriot-Watt Symposium Volume I
    R J Knops
18. Actions of fine abelian groups
    C Kosniowski
19. Closed graph theorems and webbed spaces
    M De Wilde
20. Singular perturbation techniques applied to integro-differential equations
    H Grabmüller
21. Retarded functional differential equations: A global point of view
    S E A Mohammed
22. Multiparameter spectral theory in Hilbert space
    B D Sleeman
24. Mathematical modelling techniques
    R Aris
25. Singular points of smooth mappings
    C G Gibson
26. Nonlinear evolution equations solvable by the spectral transform
    F Calogero
27. Nonlinear analysis and mechanics: Heriot-Watt Symposium Volume II
    R J Knops
28. Constructive functional analysis
    D S Bridges
29. Elongational flows: Aspects of the behaviour of model elasticoviscous fluids
    C J S Petrie
30. Nonlinear analysis and mechanics: Heriot-Watt Symposium Volume III
    R J Knops
31. Fractional calculus and integral transforms of generalized functions
    A C McBride
32. Complex manifold techniques in theoretical physics
    D E Lerner and P D Sommers
33. Hilbert's third problem: scissors congruence
    C-H Sah
34. Graph theory and combinatorics
    R J Wilson
35. The Tricomi equation with applications to the theory of plane transonic flow
    A R Manwell
36. Abstract differential equations
    S D Zaidman
37. Advances in twistor theory
    L P Hughston and R S Ward
38. Operator theory and functional analysis
    I Erdelyi
39. Nonlinear analysis and mechanics: Heriot-Watt Symposium Volume IV
    R J Knops
40. Singular systems of differential equations
    S L Campbell
41. N-dimensional crystallography
    R L E Schwarzenberger
42. Nonlinear partial differential equations in physical problems
    D Graffi
43. Shifts and periodicity for right invertible operators
    D Przeworska-Rolewicz
44. Rings with chain conditions
    A W Chatters and C R Hajarnavis
45. Moduli, deformations and classifications of compact complex manifolds
    D Sundararaman
46. Nonlinear problems of analysis in geometry and mechanics
    M Atteia, D Bancel and I Gumowski
47. Algorithmic methods in optimal control
    W A Gruver and E Sachs
48. Abstract Cauchy problems and functional differential equations
    F Kappel and W Schappacher
49. Sequence spaces
    W H Ruckle
50. Recent contributions to nonlinear partial differential equations
    H Berestycki and H Brezis
51. Subnormal operators
    J B Conway

52 Wave propagation in viscoelastic media
    F Mainardi
53 Nonlinear partial differential equations and their applications: Collège de France Seminar. Volume I
    H Brezis and J L Lions
54 Geometry of Coxeter groups
    H Hiller
55 Cusps of Gauss mappings
    T Banchoff, T Gaffney and C McCrory
56 An approach to algebraic K-theory
    A J Berrick
57 Convex analysis and optimization
    J-P Aubin and R B Vintner
58 Convex analysis with applications in the differentiation of convex functions
    J R Giles
59 Weak and variational methods for moving boundary problems
    C M Elliott and J R Ockendon
60 Nonlinear partial differential equations and their applications: Collège de France Seminar. Volume II
    H Brezis and J L Lions
61 Singular systems of differential equations II
    S L Campbell
62 Rates of convergence in the central limit theorem
    Peter Hall
63 Solution of differential equations by means of one-parameter groups
    J M Hill
64 Hankel operators on Hilbert space
    S C Power
65 Schrödinger-type operators with continuous spectra
    M S P Eastham and H Kalf
66 Recent applications of generalized inverses
    S L Campbell
67 Riesz and Fredholm theory in Banach algebra
    B A Barnes, G J Murphy, M R F Smyth and T T West
68 Evolution equations and their applications
    F Kappel and W Schappacher
69 Generalized solutions of Hamilton-Jacobi equations
    P L Lions
70 Nonlinear partial differential equations and their applications: Collège de France Seminar. Volume III
    H Brezis and J L Lions
71 Spectral theory and wave operators for the Schrödinger equation
    A M Berthier
72 Approximation of Hilbert space operators I
    D A Herrero
73 Vector valued Nevanlinna Theory
    H J W Ziegler
74 Instability, nonexistence and weighted energy methods in fluid dynamics and related theories
    B Straughan
75 Local bifurcation and symmetry
    A Vanderbauwhede
76 Clifford analysis
    F Brackx, R Delanghe and F Sommen
77 Nonlinear equivalence, reduction of PDEs to ODEs and fast convergent numerical methods
    E E Rosinger
78 Free boundary problems, theory and applications. Volume I
    A Fasano and M Primicerio
79 Free boundary problems, theory and applications. Volume II
    A Fasano and M Primicerio
80 Symplectic geometry
    A Crumeyrolle and J Grifone
81 An algorithmic analysis of a communication model with retransmission of flawed messages
    D M Lucantoni
82 Geometric games and their applications
    W H Ruckle
83 Additive groups of rings
    S Feigelstock
84 Nonlinear partial differential equations and their applications: Collège de France Seminar. Volume IV
    H Brezis and J L Lions
85 Multiplicative functionals on topological algebras
    T Husain
86 Hamilton-Jacobi equations in Hilbert spaces
    V Barbu and G Da Prato
87 Harmonic maps with symmetry, harmonic morphisms and deformations of metrics
    P Baird
88 Similarity solutions of nonlinear partial differential equations
    L Dresner
89 Contributions to nonlinear partial differential equations
    C Bardos, A Damlamian, J I Díaz and J Hernández
90 Banach and Hilbert spaces of vector-valued functions
    J Burbea and P Masani
91 Control and observation of neutral systems
    D Salamon
92 Banach bundles, Banach modules and automorphisms of C*-algebras
    M J Dupré and R M Gillette
93 Nonlinear partial differential equations and their applications: Collège de France Seminar. Volume V
    H Brezis and J L Lions
94 Computer algebra in applied mathematics: an introduction to MACSYMA
    R H Rand
95 Advances in nonlinear waves. Volume I
    L Debnath
96 FC-groups
    M J Tomkinson
97 Topics in relaxation and ellipsoidal methods
    M Akgül
98 Analogue of the group algebra for topological semigroups
    H Dzinotyiweyi
99 Stochastic functional differential equations
    S E A Mohammed

100 Optimal control of variational inequalities
   V Barbu
101 Partial differential equations and dynamical systems
   W E Fitzgibbon III
102 Approximation of Hilbert space operators. Volume II
   C Apostol, L A Fialkow, D A Herrero and D Voiculescu
103 Nondiscrete induction and iterative processes
   V Ptak and F-A Potra
104 Analytic functions – growth aspects
   O P Juneja and G P Kapoor
105 Theory of Tikhonov regularization for Fredholm equations of the first kind
   C W Groetsch
106 Nonlinear partial differential equations and free boundaries. Volume I
   J I Díaz
107 Tight and taut immersions of manifolds
   T E Cecil and P J Ryan
108 A layering method for viscous, incompressible L_p flows occupying R^n
   A Douglis and E B Fabes
109 Nonlinear partial differential equations and their applications: Collège de France Seminar. Volume VI
   H Brezis and J L Lions
110 Finite generalized quadrangles
   S E Payne and J A Thas
111 Advances in nonlinear waves. Volume II
   L Debnath
112 Topics in several complex variables
   E Ramírez de Arellano and D Sundararaman
113 Differential equations, flow invariance and applications
   N H Pavel
114 Geometrical combinatorics
   F C Holroyd and R J Wilson
115 Generators of strongly continuous semigroups
   J A van Casteren
116 Growth of algebras and Gelfand–Kirillov dimension
   G R Krause and T H Lenagan
117 Theory of bases and cones
   P K Kamthan and M Gupta
118 Linear groups and permutations
   A R Camina and E A Whelan
119 General Wiener–Hopf factorization methods
   F-O Speck
120 Free boundary problems: applications and theory, Volume III
   A Bossavit, A Damlamian and M Fremond
121 Free boundary problems: applications and theory, Volume IV
   A Bossavit, A Damlamian and M Fremond
122 Nonlinear partial differential equations and their applications: Collège de France Seminar. Volume VII
   H Brezis and J L Lions
123 Geometric methods in operator algebras
   H Araki and E G Effros
124 Infinite dimensional analysis–stochastic processes
   S Albeverio
125 Ennio de Giorgi Colloquium
   P Krée
126 Almost-periodic functions in abstract spaces
   S Zaidman
127 Nonlinear variational problems
   A Marino, L Modica, S Spagnolo and M Degiovanni
128 Second-order systems of partial differential equations in the plane
   L K Hua, W Lin and C-Q Wu
129 Asymptotics of high-order ordinary differential equations
   R B Paris and A D Wood
130 Stochastic differential equations
   R Wu
131 Differential geometry
   L A Cordero
132 Nonlinear differential equations
   J K Hale and P Martinez-Amores
133 Approximation theory and applications
   S P Singh
134 Near-rings and their links with groups
   J D P Meldrum
135 Estimating eigenvalues with *a posteriori*/*a priori* inequalities
   J R Kuttler and V G Sigillito
136 Regular semigroups as extensions
   F J Pastijn and M Petrich
137 Representations of rank one Lie groups
   D H Collingwood
138 Fractional calculus
   G F Roach and A C McBride
139 Hamilton's principle in continuum mechanics
   A Bedford
140 Numerical analysis
   D F Griffiths and G A Watson
141 Semigroups, theory and applications. Volume I
   H Brezis, M G Crandall and F Kappel
142 Distribution theorems of L-functions
   D Joyner
143 Recent developments in structured continua
   D De Kee and P Kaloni
144 Functional analysis and two-point differential operators
   J Locker
145 Numerical methods for partial differential equations
   S I Hariharan and T H Moulden
146 Completely bounded maps and dilations
   V I Paulsen
147 Harmonic analysis on the Heisenberg nilpotent Lie group
   W Schempp
148 Contributions to modern calculus of variations
   L Cesari
149 Nonlinear parabolic equations: qualitative properties of solutions
   L Boccardo and A Tesei
150 From local times to global geometry, control and physics
   K D Elworthy

151 A stochastic maximum principle for optimal control of diffusions
U G Haussmann
152 Semigroups, theory and applications. Volume II
H Brezis, M G Crandall and F Kappel
153 A general theory of integration in function spaces
P Muldowney
154 Oakland Conference on partial differential equations and applied mathematics
L R Bragg and J W Dettman
155 Contributions to nonlinear partial differential equations. Volume II
J I Díaz and P L Lions
156 Semigroups of linear operators: an introduction
A C McBride
157 Ordinary and partial differential equations
B D Sleeman and R J Jarvis
158 Hyperbolic equations
F Colombini and M K V Murthy
159 Linear topologies on a ring: an overview
J S Golan
160 Dynamical systems and bifurcation theory
M I Camacho, M J Pacifico and F Takens
161 Branched coverings and algebraic functions
M Namba
162 Perturbation bounds for matrix eigenvalues
R Bhatia
163 Defect minimization in operator equations: theory and applications
R Reemtsen
164 Multidimensional Brownian excursions and potential theory
K Burdzy
165 Viscosity solutions and optimal control
R J Elliott
166 Nonlinear partial differential equations and their applications. Collège de France Seminar. Volume VIII
H Brezis and J L Lions
167 Theory and applications of inverse problems
H Haario
168 Energy stability and convection
G P Galdi and B Straughan
169 Additive groups of rings. Volume II
S Feigelstock
170 Numerical analysis 1987
D F Griffiths and G A Watson
171 Surveys of some recent results in operator theory. Volume I
J B Conway and B B Morrel
172 Amenable Banach algebras
J-P Pier
173 Pseudo-orbits of contact forms
A Bahri
174 Poisson algebras and Poisson manifolds
K H Bhaskara and K Viswanath
175 Maximum principles and eigenvalue problems in partial differential equations
P W Schaefer
176 Mathematical analysis of nonlinear, dynamic processes
K U Grusa
177 Cordes' two-parameter spectral representation theory
D F McGhee and R H Picard
178 Equivariant K-theory for proper actions
N C Phillips
179 Elliptic operators, topology and asymptotic methods
J Roe
180 Nonlinear evolution equations
J K Engelbrecht, V E Fridman and E N Pelinovski
181 Nonlinear partial differential equations and their applications. Collège de France Seminar. Volume IX
H Brezis and J L Lions
182 Critical points at infinity in some variational problems
A Bahri
183 Recent developments in hyperbolic equations
L Cattabriga, F Colombini, M K V Murthy and S Spagnolo
184 Optimization and identification of systems governed by evolution equations on Banach space
N U Ahmed
185 Free boundary problems: theory and applications. Volume I
K H Hoffmann and J Sprekels
186 Free boundary problems: theory and applications. Volume II
K H Hoffmann and J Sprekels
187 An introduction to intersection homology theory
F Kirwan
188 Derivatives, nuclei and dimensions on the frame of torsion theories
J S Golan and H Simmons
189 Theory of reproducing kernels and its applications
S Saitoh
190 Volterra integrodifferential equations in Banach spaces and applications
G Da Prato and M Iannelli
191 Nest algebras
K R Davidson
192 Surveys of some recent results in operator theory. Volume II
J B Conway and B B Morrel
193 Nonlinear variational problems. Volume II
A Marino and M K Murthy
194 Stochastic processes with multidimensional parameter
M E Dozzi
195 Prestressed bodies
D Iesan
196 Hilbert space approach to some classical transforms
R H Picard
197 Stochastic calculus in application
J R Norris
198 Radical theory
B J Gardner
199 The C^* – algebras of a class of solvable Lie groups
X Wang

200 Stochastic analysis, path integration and dynamics
D Elworthy
201 Riemannian geometry and holonomy groups
S Salamon
202 Strong asymptotics for extremal errors and polynomials associated with Erdös type weights
D S Lubinsky
203 Optimal control of diffusion processes
V S Borkar
204 Rings, modules and radicals
B J Gardner
205 Numerical studies for nonlinear Schrödinger equations
B M Herbst and J A C Weideman
206 Distributions and analytic functions
R D Carmichael and D Mitrović
207 Semicontinuity, relaxation and integral representation in the calculus of variations
G Buttazzo
208 Recent advances in nonlinear elliptic and parabolic problems
P Bénilan, M Chipot, L Evans and M Pierre
209 Model completions, ring representations and the topology of the Pierce sheaf
A Carson
210 Retarded dynamical systems
G Stepan
211 Function spaces, differential operators and nonlinear analysis
L Paivarinta
212 Analytic function theory of one complex variable
C C Yang, Y Komatu and K Niino
213 Elements of stability of visco-elastic fluids
J Dunwoody
214 Jordan decompositions of generalised vector measures
K D Schmidt
215 A mathematical analysis of bending of plates with transverse shear deformation
C Constanda
216 Ordinary and partial differential equations Vol II
B D Sleeman and R J Jarvis
217 Hilbert modules over function algebras
R G Douglas and V I Paulsen
218 Graph colourings
R Wilson and R Nelson
219 Hardy-type inequalities
A Kufner and B Opic
220 Nonlinear partial differential equations and their applications. College de France Seminar Volume X
H Brezis and J L Lions
221 Workshop on dynamical systems
E Shiels and Z Coelho
222 Geometry and analysis in nonlinear dynamics
H W Broer and F Takens
223 Fluid dynamical aspects of combustion theory
M Onofri and A Tesei
224 Approximation of Hilbert space operators. Volume I. 2nd edition
D Herrero
225 Surveys of some recent results in operator theory Volume III
J B Conway and B B Morrel
226 Local cohomology and localization
J L Bueso Montero, B Torrecillas Jover and A Verschoren
227 Sobolev spaces of holomorphic functions
F Beatrous and J Burbea
228 Numerical analysis. Volume III
D F Griffiths and G A Watson
229 Recent developments in structured continua. Volume III
D De Kee and P Kaloni
230 Boolean methods in interpolation and approximation
F J Delvos and W Schempp

A mathematical analysis of bending of plates with transverse shear deformation

For Lia

C Constanda
University of Strathclyde

A mathematical analysis of bending of plates with transverse shear deformation

Copublished in the United States with
John Wiley & Sons, Inc., New York

Longman Scientific & Technical,
Longman Group UK Limited,
Longman House, Burnt Mill, Harlow
Essex CM20 2JE, England
and Associated Companies throughout the world.

Copublished in the United States with
John Wiley & Sons, Inc., 605 Third Avenue, New York, NY 10158

© C Constanda 1990

All rights reserved; no part of this publication
may be reproduced, stored in a retrieval system,
or transmitted in any form or by any means, electronic,
mechanical, photocopying, recording, or otherwise,
without either the prior written permission of the Publishers
or a licence permitting restricted copying in the United Kingdom
issued by the Copyright Licensing Agency Ltd,
33-34 Alfred Place, London, WC1E 7DP.

First published 1990

AMS Subject Classification: (Main) 35J55, 73C05, 73K10
                                (Subsidiary) 45F15, 35C15, 42C15

ISSN 0269-3674

British Library Cataloguing in Publication Data
Constanda, C
  A mathematical analysis of bending of plates with transverse shear deformation.
  1. Structures. Plates. Applied mechanics. Mathematics
  I. Title
  624.1'7765'0151

ISBN 0-582-04044-2

Library of Congress Cataloging-in-Publication Data
Constanda, C. (Christian)
  A mathematical analysis of bending of plates with transverse shear deformation.
  p. cm.– (Pitman research notes in mathematics series, 0269-3674 215)
  Bibliography: p.
  ISBN 0-470-21305-1
  1. Plates (Engineering)—Mathematical models. 2. Flexure.
  3.Shear (Mechanics) 4. Mathematical analysis. I. Title.
  II. Series.
TA660.P6C86 1989
624,1'7765'0151535—dc20   89-8076
                                      CIP

Printed and bound in Great Britain
by Biddles Ltd, Guildford and King's Lynn

Contents

Preface

1. Singular integral equations ... 1
    1.1. Introduction ... 1
    1.2. Geometry of the boundary curve ... 4
    1.3. Properties of the boundary layer ... 8
    1.4. Integrals with singular kernels ... 18
    1.5. The harmonic potentials ... 35
    1.6. Other potential-type functions ... 44
    1.7. Complex singular kernels ... 54
    1.8. Singular integral equations ... 62

2. Bending of elastic plates ... 73
    2.1. The two-dimensional plate model ... 73
    2.2. Singular solutions ... 80
    2.3. The case of the exterior domain ... 85
    2.4. Uniqueness of regular solutions ... 88
    2.5. Elastic potentials with smooth densities ... 90
    2.6. Elastic potentials with integrable densities ... 102
    2.7. Existence of regular solutions ... 110
    2.8. Smoothness of the integrable solutions ... 121

3. Complex variable treatment ... 124
    3.1. Complex representation of the stresses ... 124
    3.2. The traction boundary value problem ... 128
    3.3. The displacement boundary value problem ... 129
    3.4. Arbitrariness in the complex potentials ... 134
    3.5. Bounded multiply connected domain ... 135

3.6. Unbounded multiply connected domain	137
3.7. Example	141
3.8. Physical significance of the restrictions	142
4. Generalized Fourier series	144
4.1. The interior Dirichlet problem	144
4.2. The interior Neumann problem	150
4.3. The exterior Dirichlet problem	156
4.4. The exterior Neumann problem	163
4.5. Numerical example	164
References	166

Preface

Approximate theories of bending of thin elastic plates have been around since the middle of the last century. The reason for their existence is twofold: on the one hand, they reduce the full three-dimensional model to a simpler one in only two idependent variables; on the other, they give prominence to the main characteristics of bending, neglecting other effects that are of little interest in the study of this physical process.

In spite of their good agreement with experiments and their wide use by engineers in practical applications, such theories never acquire true legitimacy until they have been subjected to the customary ordeal by rigorous mathematical analysis and pronounced valid. The classical (Kirchhoff) model has been studied almost completely (see, for example, [3] and [20]). In this book we aim to do the same for the modern ones (Reissner, Mindlin), discussing the existence, uniqueness and approximation of their regular solutions by means of the boundary integral equation and stress function methods.

With the exception of a few results, which are quoted without proof, the presentation is self-contained. To avoid discouraging the non-specialist reader, the use of the language of functional analysis has been kept to a minimum.

Here is a brief account of the contents of the book.

Chapter 1 prepares the ground for the investigation of the equations of bending by developing a number of prerequisites concerning the behaviour of integrals with singular kernels near the boundary of the domain.

In Chapter 2 we formulate and solve the interior and exterior displacement and traction boundary value problems by means of elastic single and double layer potentials.

Chapter 3 is devoted to the construction of the complete integral of the system of equilibrium equations and the elucidation of the physical meaning of certain analytic constraints imposed earlier on the asymptotic behaviour of the solutions.

Finally, in Chapter 4 we indicate how the method of generalized Fourier series can be adapted to provide approximate solutions for the Dirichlet and Neumann problems.

Part of the results incorporated in this book have been published in [7]–[19]. The technique developed in Chapters 1 and 2 has also been extended to the case of bending of micropolar plates in [5] and [35]–[37].

Before proceeding with the business of mathematical analysis, I would like to thank a number of people who, in some way or other, helped me in the process of preparation of this book: Bill Coberly and the rest of his team in the Department of Mathematical and Computer Sciences at the University of Tulsa, Oklahoma, who created such a pleasant and stimulating atmosphere for me during my sabbatical there that the idea of writing this book became irresistible; Gary Roach, for assistance with the preliminary stages; Peter Schiavone, for playing an enthusiastic devil's advocate on numerous occasions and asking some awkward questions, the answers to which generated several sections of the first two chapters; Rolf Leis and Derick Atkinson, for inviting me to visit respectively the University of Bonn and the University of Toronto, where certain segments of the work were completed; Dan Constanda, for his hot pursuit of the numbers in the final section; my wife, for unflagging moral support and tremendous patience; Des McGhee, the 'friend in need' when I was short of a microcomputer; and the staff of Longman, who allowed me free rein in the graphical arrangements of the camera-ready copy and were so understanding over deadlines.

Lastly, I cannot help giving myself a pat on the back for turning out to be such an accomplished TeX user.

1 Singular integral equations

1.1. Introduction

Throughout the book we make use of a number of well-established symbols and conventions. Thus, Greek and Latin subscripts take the values 1, 2 and 1, 2, 3, respectively, summation over repeated indices is understood, $x = (x_1, x_2)$ and $x = (x_1, x_2, x_3)$ are generic points referred to orthogonal Cartesian coordinates in $\mathbf{R}^2$ and $\mathbf{R}^3$, a superscript T indicates matrix transposition, $(\ldots)_{,\alpha} = \partial(\ldots)/\partial x_\alpha$, Δ is the Laplacian, and δ_{ij} are the Kronecker delta. Other notation will be defined as it occurs in the text.

The elastostatic behaviour of a three-dimensional homogeneous and isotropic body is described by the equilibrium equations

$$t_{ij,j} + f_i = 0 \qquad (1.1)$$

and the constitutive relations

$$t_{ij} = \lambda u_{k,k} \delta_{ij} + \mu(u_{i,j} + u_{j,i}) \qquad (1.2)$$

(see, for example, [21]). Here $t_{ij} = t_{ji}$ are the internal stresses, u_i the displacements, f_i the body forces, and λ and μ the Lamé constants of the material.

The components of the resultant stress vector t in a direction $n = (n_1, n_2, n_3)^T$ are

$$t_i = t_{ij} n_j, \qquad (1.3)$$

and the internal energy per unit volume is

$$\mathcal{E} = \tfrac{1}{4} t_{ij}(u_{i,j} + u_{j,i}) = \tfrac{1}{2} t_{ij} u_{i,j}. \qquad (1.4)$$

A *thin plate* is an elastic body that occupies a region $\bar{S} \times [-\frac{1}{2}h_0, \frac{1}{2}h_0]$ in $\mathbf{R}^3$, where S is a domain in $\mathbf{R}^2$ and $0 < h_0 = \text{const} \ll \text{diam } S$ is the thickness. The special form of such a body suggests that in the study of its small deformations certain simplifying assumptions may be introduced, which lead to two-dimensional theories that are easier to handle but still describe adequately the salient features of the deformation state. In what follows we are concerned exclusively with the process of bending.

The first truly systematic theory of bending of thin elastic plates was proposed by Kirchhoff [23]. Under his assumptions the displacement field becomes

$$u_\alpha = -x_3 u_{3,\alpha},$$
$$u_3 = u_3(x_\gamma), \qquad (1.5)$$

and from (1.1) and (1.2) it follows that

$$\Delta\Delta u_3 = \frac{p}{D},$$

where p is the resultant load on the faces $x_3 = \pm\frac{1}{2}h_0$ of the plate and $D = h_0^3 \mu(\lambda+\mu)[3(\lambda+2\mu)]^{-1}$ is the rigidity modulus. This theory, though producing good approximations in many practical cases, neglects completely the effects of the transverse shear forces since (1.2) and (1.5) yield $t_{3\alpha} = 0$ throughout the plate. It also gives rise to a few mathematical discrepancies: certain stress components are neglected in some equations but not in others. In addition, the unknown deflection u_3 can satisfy only two boundary conditions on ∂S instead of the physically expected three.

Reissner (see [30]–[33]) takes transverse shear into account by assuming that

$$t_{\alpha\beta} = \frac{h_0^2}{12} x_3 M_{\alpha\beta}(x_\gamma),$$

$$t_{\alpha 3} = \frac{3}{2h_0}\left[1 - \left(\frac{2}{h_0}\right)^2 x_3^2\right] Q_\alpha(x_\gamma),$$

and uses the principle of least work to derive a sixth order theory that accommodates three boundary conditions. While this is a more complete model than Kirchhoff's, it does not deliver the expression of the displacements but only that of their averages.

Hencky [22], Bollé [2], Uflyand [39], and Mindlin [25] introduce the effects of transverse shear deformation in a somewhat different manner. More precisely, they start with the displacement assumption

$$u_\alpha = x_3 v_\alpha(x_\gamma),$$
$$u_3 = v_3(x_\gamma) \tag{1.6}$$

and arrive at the equations of an approximate sixth order theory by averaging (1.1) and (1.2) over the thickness of the plate. As in the case of Reissner's, these equations allow three conditions to be prescribed on the boundary. Unfortunately, they suffer from the same lack of rigour, due to the fact that t_{33} is neglected in the constitutive relations, which also contain so-called correction factors.

The above theories have subsequently been refined in various ways, but all their versions pursue the same ultimate goal: to offer as much valid information as possible on the characteristics of bending, while at the same time reducing the problem to a simpler one in two dimensions (see [34] for a concise survey of this topic).

Here we are not concerned with the advantages of one theory over another from a physical standpoint, but with their mathematical treatment. As the model of our analysis we choose an approximation based solely on the kinematic assumption (1.6), thus avoiding inconsistencies that might otherwise be introduced through over-simplification. However, our technique is equally applicable—with very little modification regarding the coefficients—to all existing sixth order theories where the system of equilibrium equations is elliptic.

1.2. Geometry of the boundary curve

For simplicity, we use the same symbol to indicate both a point and its position vector in $\mathbf{R}^2$. Also, vector functions are not distinguished from scalar ones by any special marks, their nature being obvious from the context.

Let the boundary ∂S of S be a simple closed curve of length $|\partial S|$, whose equation in terms of its arc coordinate is

$$x = \psi(s), \quad s \in [0, |\partial S|], \quad \psi(0) = \psi(|\partial S|),$$

with the inverse relationship written as $s = s(x)$, $x \in \partial S$.

Throughout what follows we assume that ∂S is a C^2-curve, that is, ψ is twice continuously differentiable on $[0, |\partial S|]$ and

$$\frac{d\psi}{ds}(0+) = \frac{d\psi}{ds}(|\partial S|-),$$
$$\frac{d^2\psi}{ds^2}(0+) = \frac{d^2\psi}{ds^2}(|\partial S|-).$$

Denoting by $\nu(x)$ the unit outward normal to ∂S at $x \in \partial S$ and by $\tau(x)$ the unit tangent at x orientated in the direction in which s increases, chosen so that $\{\nu(x), \tau(x)\}$ is right-handed, we can write

$$\tau_\alpha = \varepsilon_{\beta\alpha}\nu_\beta, \tag{1.7}$$

where $\varepsilon_{\alpha\beta}$ is the alternating symbol.

If $\kappa(x)$ is the algebraic value of the curvature at $x \in \partial S$, then

$$\frac{d}{ds}\tau(x) = -\kappa(x)\nu(x),$$
$$\frac{d}{ds}\nu(x) = \kappa(x)\tau(x). \tag{1.8}$$

Let $\langle \cdot, \cdot \rangle$ be the standard inner product and $|x|^2 = x_1^2 + x_2^2$ the Euclidean norm in $\mathbf{R}^2$.

1.1. Lemma. *There is a constant $q > 0$ such that*

$$|\langle \nu(x), x - y \rangle| \leq q|x - y|^2, \tag{1.9}$$

$$|\nu(x) - \nu(y)| \leq q|x - y| \tag{1.10}$$

for all $x, y \in \partial S$.

Proof. Since ∂S is a C^2-curve, we can define

$$\kappa_0 = \sup_{x \in \partial S} |\kappa(x)|. \tag{1.11}$$

It is obvious that $\kappa_0 > 0$, for $\kappa_0 = 0$ would imply that ∂S were a straight line, therefore, not a closed curve.

Let $x = \psi(s)$ and $y = \psi(t)$. By (1.8),

$$\frac{\partial}{\partial s(x)} |x - y|^2 = 2\langle \tau(x), x - y \rangle,$$

$$\frac{\partial^2}{\partial s^2(x)} |x - y|^2 = 2\big[1 - \kappa(x)\langle \nu(x), x - y \rangle\big],$$

therefore, for $y \in \partial S$ sufficiently close to x

$$|x - y|^2 = \big[1 - \kappa(x')\langle \nu(x'), x' - y \rangle\big](s - t)^2,$$

$$\langle \nu(y), x - y \rangle = -\tfrac{1}{2}\kappa(x'')\langle \nu(x''), \nu(y) \rangle (s - t)^2,$$

where $x', x'' \in \partial S$ lie between x and y.

First suppose that $|x - y| \leq (2\kappa_0)^{-1}$. Then

$$|x - y|^2 \geq \tfrac{1}{2}(s - t)^2,$$

consequently,
$$|\langle \nu(y), x - y \rangle| \leq \kappa_0 |x - y|^2.$$

For $|x - y| > (2\kappa_0)^{-1}$ we have
$$|\langle \nu(y), x - y \rangle| \leq |x - y| \leq |\partial S| < 4\kappa_0^2 |\partial S| \, |x - y|^2.$$

Hence, for all $x, y \in \partial S$
$$|\langle \nu(y), x - y \rangle| \leq \max\{\kappa_0, 4\kappa_0^2 |\partial S|\} |x - y|^2.$$

Next,
$$\nu_\alpha(x) - \nu_\alpha(y) = \kappa(x''') \tau_\alpha(x''')(s - t),$$

where $x''' \in \partial S$ lies between x and y. Reasoning as above, we find that for all $x, y \in \partial S$
$$|\nu(x) - \nu(y)| \leq 8\kappa_0 |x - y|.$$

The inequalities (1.9) and (1.10) are now obtained by setting
$$q = \max\{8\kappa_0, 4\kappa_0^2 |\partial S|\}.$$

1.2. Lemma. *Let $x, y \in \partial S$, and let α be the angle between $\nu(x)$ and $\nu(y)$ and γ the acute angle between the support lines of $\nu(x)$ and $x - y$. If $0 < r = \mathrm{const} \leq (2q)^{-1}$, where q is the constant specified in Lemma 1.1, then*

(i) $\frac{1}{2} \leq \cos\alpha \leq 1$,

(ii) $\frac{1}{2} \leq \sin\gamma \leq 1$

for all x and y such that $|x - y| \leq r$.

Proof. (i) By Lemma 1.1,
$$\langle \nu(x), \nu(y) \rangle = 1 - \langle \nu(x), \nu(x) - \nu(y) \rangle \geq 1 - qr \geq \tfrac{1}{2}.$$

(ii) By the Mean Value Theorem, there is an $x' \in \partial S$ between x and y such that $\tau(x')$ is parallel to $x - y$. By (i), the acute angle β between $\tau(x)$ and $x - x'$ satisfies $\frac{1}{2} \leq \cos\beta \leq 1$. The statement (ii) now follows from the fact that $\sin\gamma = \cos\beta$.

1.3. Lemma. *If*

$$\Sigma_{x,r} = \{y \in \partial S : |x - y| \leq r\}, \quad x \in \partial S, \tag{1.12}$$

with r as in Lemma 1.2, then

$$\tfrac{1}{2}|s - t| \leq |x - y| \leq |s - t| \tag{1.13}$$

for all $x \in \partial S$ and $y \in \Sigma_{x,r}$, where $x = \psi(s)$ and $y = \psi(t)$.

Proof. Let $a = \psi(s_1)$ and $b = \psi(s_2)$ be the end-points of $\Sigma_{x,r}$. Without loss of generality we may assume that $0 \leq s_1 < s_2$. For any $y \in \Sigma_{x,r}$ between x and b we have

$$\frac{d}{dt}|x - y| = -\frac{\langle \tau(y), x - y \rangle}{|x - y|} = \cos\beta(t),$$

where β is the acute angle between the support lines of $\tau(y)$ and $x - y$. Hence,

$$|x - y| = \int_s^t \cos\beta(\sigma)\,d\sigma.$$

By the Mean Value Theorem,

$$|x - y| = (s - t)\cos\beta(\eta),$$

with $\eta \in \partial S$ between x and y. Similarly, for any $y \in \Sigma_{x,r}$ between a and x

$$\frac{d}{dt}|x - y| = -\cos\beta(t),$$

so that, in general, for any $y \in \Sigma_{x,r}$ we can write

$$|x - y| = c(s, t)|s - t|,$$

where, by Lemma 1.2, $\frac{1}{2} \leq c(s, t) \leq 1$.

1.4. Remark. From the proof of Lemma 1.3 it is clear that $|x - y|$ is a monotonic function of t on the intervals $\{t : y \in \Sigma_{x,r}, t \leq s\}$ and $\{t : y \in \Sigma_{x,r}, t \geq s\}$, decreasing on the former and increasing on the latter. This implies that $|x - y'| \neq |x - y''|$ for all $y', y'' \in \Sigma_{x,r}, y' \neq y''$, and that there is a bijective correspondence between the points of $\Sigma_{x,r}$ and those of its projection on the tangent to ∂S at x.

1.5. Remark. A slightly modified pair of inequalities (1.13) holds for all $x, y \in \partial S$ if by $|s - t|$ we understand the length of the shorter arc of ∂S joining x and y. Since for $|x - y| > r$

$$|s - t| \leq |\partial S| \leq \frac{|\partial S|}{r}|x - y|,$$

we conclude that for all $x, y \in \partial S$

$$c|s - t| \leq |x - y| \leq |s - t|,$$

where $c = \min\{\frac{1}{2}, r/|\partial S|\}$.

1.3. Properties of the boundary layer

Many of the results in this book are proved by considering the behaviour of certain two-point functions in the neighbourhood of the boundary. To help the fluency of such proofs, here we make a preliminary examination of some frequently used properties.

1.6. Lemma. *The curves*

$$\partial S_\sigma = \{x \in \mathbf{R}^2 : x = \xi + \sigma\nu(\xi),\ \xi \in \partial S\},$$
$$\sigma = \text{const},\quad 0 < |\sigma| < \kappa_0^{-1},$$

where κ_0 is given by (1.11), are parallel to ∂S and well defined, that is, the support lines of $\nu(\xi)$ and $\nu(\xi')$ do not intersect between $\partial S_{-1/\kappa_0}$ and $\partial S_{1/\kappa_0}$ for any $\xi, \xi' \in \partial S$.

Proof. Let ϖ be the arc coordinate on ∂S_σ. Then

$$\tau(x) = \frac{dx}{d\varpi} = \left(\frac{d\xi}{ds} + \sigma\frac{d\nu}{ds}\right)\frac{ds}{d\varpi} = [1 + \sigma\kappa(\xi)]\tau(\xi)\frac{ds}{d\varpi}.$$

Since

$$(d\varpi)^2 = |dx|^2 = [1 + \sigma\kappa(\xi)](ds)^2$$

and

$$1 + \sigma\kappa(\xi) \geq 1 - |\sigma|\kappa_0 > 0,$$

it follows that

$$\tau(x) = \tau(\xi).$$

If for some x we have

$$x = \xi + \sigma\nu(\xi) = \xi' + \sigma'\nu(\xi'),\quad |\sigma|, |\sigma'| < \kappa_0^{-1},\quad \xi \neq \xi',$$

then $\tau(x) = \tau(\xi) = \tau(\xi')$. This leads to $\nu(\xi) = \nu(\xi')$, which contradicts the above assumption.

1.7. Lemma. *Consider the boundary layer*

$$S_{\sigma_0} = \{x \in \mathbf{R}^2 : x = \xi + \sigma\nu(\xi),\ \xi \in \partial S,\ |\sigma| \leq \sigma_0\}.$$

If x, $x' \in S_{r/4}$, $|x-x'| < \frac{1}{4}r$, $x' = \xi' + \sigma'\nu(\xi')$, and $r \leq \min\{\kappa_0^{-1}, (2q)^{-1}\}$, where κ_0 is defined by (1.11) and q is the number specified in Lemma 1.1, then

$$|\xi - \xi'| < 4|x - x'|.$$

Proof. Without loss of generality suppose that $|x - \xi| \geq |x' - \xi'|$. Let ξ_0 be the point of intersection of the support lines of $\nu(\xi)$ and $\nu(\xi')$, and η the point on the line through ξ and ξ_0 such that $\eta - x'$ is parallel to $\xi - \xi'$. By Lemma 1.6, since ξ_0 does not lie between $\partial S_{-1/\kappa_0}$ and $\partial S_{1/\kappa_0}$, we have

$$|\xi_0 - \xi| > \frac{1}{2\kappa_0},$$

therefore,

$$\frac{|\eta - x'|}{|\xi - \xi'|} \geq 1 - \frac{|\eta - \xi|}{|\xi_0 - \xi|} > 1 - \frac{r/4}{1/(2\kappa_0)} = 1 - \frac{1}{2}\kappa_0 r \geq \frac{1}{2}. \quad (1.14)$$

Let ϑ be the angle between $\xi - \xi_0$ and $x' - x$, and γ the acute angle between $\nu(\xi)$ and $\xi' - \xi$. By (1.14),

$$|\xi - \xi'| \leq 2|\eta - x'| \leq 2(|x - x'| + |\eta - x|) < 2(\tfrac{1}{4}r + \tfrac{1}{4}r) = r,$$

so, by Lemma 1.2 and (1.14),

$$|x - x'| = \frac{\sin\gamma}{\sin\vartheta}|\eta - x'| \geq \tfrac{1}{2}|\eta - x'| > \tfrac{1}{4}|\xi - \xi'|,$$

as required.

1.8. Lemma. *With the notation in Lemma 1.7, if x, $x' \in S_{r/4}$ satisfy $|x - x'| < \frac{1}{8}r$, $r \leq \min\{\kappa_0^{-1}, (2q)^{-1}\}$, $\xi = \psi(s)$, $\xi' = \psi(s')$, $y = \psi(t)$, and*

$$\Sigma_1 = \{y \in \Sigma_{\xi,r} : |s - t| \leq 8|x - x'|\}, \quad (1.15)$$

where $\Sigma_{\xi,r}$ is defined by (1.12), then $\xi' \in \Sigma_1$ and for all $y \in \Sigma_1$

(i) $|x-y| \geq \frac{1}{2}|\xi - y|$;

(ii) $|x-y| \geq \frac{1}{2}|x - \xi|$;

(iii) $|x'-y| \geq \frac{1}{2}|\xi' - y| \geq \frac{1}{4}|s'-t|$.

Proof. By Lemma 1.7,
$$|\xi - \xi'| < 4|x - x'| < r,$$
so, by Lemmas 1.3 and 1.7,
$$|s - s'| \leq 2|\xi - \xi'| < 8|x - x'|,$$
which implies that $\xi' \in \Sigma_1$.

(i) Let γ be the acute angle between $\nu(\xi)$ and $\xi - y$, and ϑ the angle between $x - \xi$ and $x - y$. Since $|\xi - y| \leq r$, Lemma 1.2 yields
$$|x - y| = \frac{\sin \gamma}{\sin \vartheta}|\xi - y| \geq \frac{1}{2}|\xi - y|.$$

(ii) As above,
$$|x - y| = \frac{\sin \gamma}{\sin(\gamma + \vartheta)}|x - \xi| \geq \frac{1}{2}|x - \xi|.$$

(iii) The fact that $\xi' \in \Sigma_1$ implies that
$$|\xi' - y| \leq |s' - t| \leq 8|x - x'| < r.$$

Repeating the argument in (i) with ξ' instead of ξ, we obtain
$$|x' - y| \geq \frac{1}{2}|\xi' - y|.$$

Also, by Lemma 1.3,
$$|\xi' - y| \geq \frac{1}{2}|s' - t|,$$

which completes the proof.

1.9. Lemma. *With the notation in Lemma 1.8, if $x, x' \in S_{r/4}$ satisfy $|x - x'| < \frac{1}{8}r$, $r \leq \min\{\kappa_0^{-1}, (2q)^{-1}\}$, and*

$$\Sigma_2 = \Sigma_{\xi,r} \setminus \Sigma_1 = \{y \in \Sigma_{\xi,r} : |s - t| > 8|x - x'|\}, \qquad (1.16)$$

then $\operatorname{mes} \Sigma_2 > 0$, *and for all* $y \in \Sigma_2$

(i) $|x - y| \geq \frac{1}{2}|\xi - y|$;

(ii) $|x' - y| \geq \frac{1}{4}|\xi - y|$;

(iii) $|x - x'| < \frac{1}{2}|x - y|$;

(iv) $|\xi' - y| < 3|\xi - y|$.

Proof. Since $\operatorname{mes} \Sigma_1 \leq 16|x - x'| < 2r \leq \operatorname{mes} \Sigma$, it follows immediately that $\operatorname{mes} \Sigma_2 > 0$.

(i) This is proved exactly as the first assertion of Lemma 1.8.

(ii) By Lemma 1.3 and (i),

$$|x' - y| \geq |x - y| - |x - x'| \geq |x - y| - \tfrac{1}{8}|s - t|$$
$$\geq \tfrac{1}{2}|\xi - y| - \tfrac{1}{4}|\xi - y| = \tfrac{1}{4}|\xi - y|.$$

(iii) By Lemma 1.7,

$$|\xi - \xi'| < 4|x - x'| < r.$$

Applying Lemma 1.3 and (i), we now obtain

$$|x - x'| < \tfrac{1}{8}|s - t| \leq \tfrac{1}{4}|\xi - y| \leq \tfrac{1}{2}|x - y|.$$

(iv) By Lemma 1.8, $\xi' \in \Sigma_1$. Consequently, by Lemma 1.3,

$$|\xi' - y| \leq |\xi - y| + |\xi - \xi'| \leq |\xi - y| + |s - s'|$$
$$< |\xi - y| + |s - t| \leq 3|\xi - y|,$$

as required.

1.10. Lemma. *With the notation in Lemma 1.8, if x, $x' \in S_{r/4}$ satisfy $|x - x'| < \frac{1}{4}r$, $r \leq \min\{\kappa_0^{-1}, (2q)^{-1}\}$, and $\Sigma_{\xi,r}$ is defined by (1.12), then for all $y \in \partial S \setminus \Sigma_{\xi,r}$*

(i) $|x - x'| < \frac{1}{2}|x - y|$;

(ii) $|x - y| > \frac{3}{4}|\xi - y|$;

(iii) $|x' - y| > \frac{1}{4}|\xi - y|$;

(iv) $|\xi' - y| < 2|\xi - y|$.

Proof. Let $y \in \partial S \setminus \Sigma_{\xi,r}$.

(i) Since $x \in S_{r/4}$,

$$|x - y| \geq |\xi - y| - |x - \xi| > r - \tfrac{1}{4}r = \tfrac{3}{4}r > 2|x - x'|.$$

(ii) As above,

$$|x - y| \geq |\xi - y| - \tfrac{1}{4}r > |\xi - y| - \tfrac{1}{4}|\xi - y| = \tfrac{3}{4}|\xi - y|.$$

(iii) By (ii),

$$|x' - y| \geq |x - y| - |x - x'| > |x - y| - \tfrac{1}{2}r$$
$$> \tfrac{3}{4}|\xi - y| - \tfrac{1}{2}|\xi - y| = \tfrac{1}{4}|\xi - y|.$$

(iv) By Lemma 1.7,
$$|\xi' - y| \le |\xi - y| + |\xi - \xi'| < |\xi - y| + 4|x - x'|$$
$$< |\xi - y| + r < 2|\xi - y|.$$

1.11. Lemma. *With the notation in Lemmas 1.8 and 1.9, if $x, x' \in S_{r/4}$ satisfy $0 < |x - x'| < \frac{1}{8}r$, $r \le \min\{\frac{1}{2}, \kappa_0^{-1}, (2q)^{-1}\}$, then there are constants $c_1, c_2, c_3, c_4 > 0$ such that*

(i) $\int_{\Sigma_1} |x-y|^{-\gamma} ds(y) \le c_1 |x-x'|^{1-\gamma}$ *for any $\gamma < 1$, where c_1 depends on γ;*

(ii) $\int_{\Sigma_1} |x'-y|^{-\gamma} ds(y) \le c_2 |x-x'|^{1-\gamma}$ *for any $\gamma < 1$, where c_2 depends on γ;*

(iii) $\int_{\Sigma_2} |x-y|^{-\gamma-1} ds(y) \le c_3 |x-x'|^{-\gamma}$ *for any $\gamma \in (0,1)$, where c_3 depends on γ;*

(iv) $\int_{\Sigma_2} |x-y|^{-1} ds(y) \le c_4 |\ln|x-x'||$.

Proof. Let $\delta = |x - x'|$, $x = \xi + \sigma\nu(\xi)$, $x' = \xi' + \sigma'\nu(\xi')$, $\xi, \xi' \in \partial S$, $\xi = \psi(s)$, $\xi' = \psi(s')$, $y = \psi(t)$, and

$$\Gamma_1 = \{t : y \in \Sigma_1\} = \{t : |s-t| \le 8\delta\},$$
$$\Gamma_2 = \{t : y \in \Sigma_2\} = \{t : y \in \Sigma_{\xi,r}, |s-t| > 8\delta\}. \quad (1.17)$$

Without loss of generality suppose that the point corresponding to the origin of the arc coordinate lies outside $\Sigma_{\xi,r}$.

(i) By Lemmas 1.8 and 1.3,

$$\int_{\Sigma_1} |x-y|^{-\gamma} ds(y) \le 2^\gamma \int_{\Sigma_1} |\xi-y|^{-\gamma} ds(y) \le 4^\gamma \int_{\Gamma_1} |s-t|^{-\gamma} dt$$
$$= \frac{2 \cdot 4^\gamma \cdot 8^{1-\gamma}}{1-\gamma} \delta^{1-\gamma}.$$

(ii) By Lemma 1.8 and the definition of Γ_1 in (1.17),
$$s' \in \Gamma_1,$$
which means that $|s - s'| \leq 8\delta$. Therefore, as above,
$$\int_{\Sigma_1} |x' - y|^{-\gamma} ds(y) \leq 4^\gamma \int_{\Gamma_1} |s' - t|^{-\gamma} dt \leq \frac{2 \cdot 4^\gamma}{1 - \gamma}(|s - s'| + 8\delta)^{1-\gamma}$$
$$\leq \frac{2 \cdot 4^\gamma \cdot 16^{1-\gamma}}{1 - \gamma} \delta^{1-\gamma}.$$

(iii) Applying Lemmas 1.9 and 1.3 and calculating the integral explicitly in terms of the end-points of Γ_2, we obtain
$$\int_{\Sigma_2} |x - y|^{-\gamma - 1} ds(y) \leq 4^{\gamma+1} \int_{\Gamma_2} |s - t|^{-\gamma - 1} dt \leq \frac{8}{\gamma} \delta^{-\gamma}.$$

(iv) If a and b are the end-points of $\Sigma_{\xi,r}$, $a = \psi(s_a)$, $b = \psi(s_b)$, $s_a < s < s_b$, then
$$\int_{\Sigma_2} |x - y|^{-1} ds(y) \leq 4 \int_{\Gamma_2} |s - t|^{-1} dt$$
$$= 4\big[\ln(s - s_a) + \ln(s - s_b) - 2\ln(8\delta)\big].$$

Since $r < \frac{1}{2}$, from Lemma 1.3 it follows that
$$s - s_a \leq 2|x - a| = 2r < 1,$$
and, similarly, $s_b - s < 1$. Hence,
$$\int_{\Sigma_2} |x - y|^{-1} ds(y) < 8|\ln \delta|,$$
which completes the proof.

1.12. Remark. It is obvious that all the conditions in Lemmas 1.2, 1.3, 1.6, 1.10, and 1.11 are satisfied if, for example, we choose $x, x' \in S_{r/4}$ such that
$$|x - x'| < \tfrac{1}{8}r, \quad r = \min\{\tfrac{1}{2}, \kappa_0^{-1}, (2q)^{-1}\}. \tag{1.18}$$

From now on we use the notation
$$S_0 = S_{r/4},$$
$$S_0^+ = \{x \in S_0 : x = \xi + \sigma\nu(\xi), \ \xi \in \partial S, \ -\tfrac{1}{4}r \leq \sigma < 0\},$$
$$S_0^- = S_0 \setminus \bar{S}_0^+.$$

1.13. Remark. For $x \in \partial S$ we introduce local coordinates (ρ, ω) along the positive tangent and inward normal to ∂S at x, respectively. Since ∂S is a simple C^2-curve, in accordance with Remark 1.4 there is a function f twice continuously differentiable on $[-r, r]$ satisfying $f(0) = f'(0) = 0$ and such that the equation of $\Sigma_{x,r}$ can be written in the form $\omega = f(\rho)$. If α is the angle between $\nu(x)$ and $\nu(y)$ and β the acute angle between the support lines of $x - y$ and $\tau(x)$, then, by Lemma 1.2, we find that for all $y \in \Sigma_{x,r}$
$$|f(\rho)| = |x - y|\sin\beta \leq \frac{\sqrt{3}}{2}r,$$
$$|f'(\rho)| = |\tan\alpha| \leq \sqrt{3},$$
$$|f''(\rho)| \leq \kappa_0\left[1 + f'^2(\rho)\right]^{3/2} \leq \bar{\kappa}_0.$$

1.14. Theorem. *If* $x, y \in \partial S$, $x = \psi(s)$ *and* $y = \psi(t)$, *then*
$$\frac{|x - y|}{|s - t|} \to 1 \quad \text{as} \quad y \to x, \quad \text{uniformly on } \partial S.$$

Proof. Using the local coordinates (ρ, ω) defined in Remark 1.13, we can write
$$f(\rho) = \tfrac{1}{2}\rho^2 f''(\rho_1), \quad f'(\rho) = \rho f''(\rho_2), \tag{1.19}$$

with ρ_1 and ρ_2 between 0 and ρ. Consequently, for $\rho > 0$

$$f_1(\rho) = |x - y| = [\rho^2 + f^2(\rho)]^{1/2} = \rho + \tfrac{1}{2}\rho^2 f_1''(\rho'),$$
$$f_2(\rho) = |s - t| = \int_0^\rho [1 + f'^2(\theta)]\, d\theta = \rho + \tfrac{1}{2}\rho^2 f_2''(\rho''), \qquad (1.20)$$

where $0 < \rho', \rho'' < \rho$. From this we find that

$$\left|\frac{f_1(\rho)}{f_2(\rho)} - 1\right| = \frac{\rho|f_1''(\rho') - f_2''(\rho'')|}{|2 + \rho f_2''(\rho'')|}.$$

From (1.19) and (1.20) we have

$$f_1''(\rho) = \rho\big[\tfrac{1}{4}f''^2(\rho_1) + f''^2(\rho_2) + \tfrac{1}{2}f''(\rho_1)f''(\rho) - f''(\rho_1)f''(\rho_2)$$
$$+ \tfrac{1}{8}\rho^2 f''^3(\rho_1)f''(\rho)\big]\big[1 + \tfrac{1}{4}\rho^2 f''^2(\rho_1)\big]^{-3/2},$$
$$f_2''(\rho) = f'(\rho)f''(\rho)\big[1 + f'^2(\rho)\big]^{-1/2},$$

so that, by Remark 1.13,

$$|f_1''(\rho)| \le c_1, \quad |f_2''(\rho)| \le c_1, \quad c_1 = \text{const} > 0,$$

for all $\rho \in [-r, r]$ and all $x \in \partial S$.

Let $0 < \rho < c_1^{-1}$. Since

$$|2 + \rho f_2''(\rho'')| \ge 2 - \rho|f_2''(\rho'')| > 1,$$

we conclude that

$$\left|\frac{f_1(\rho)}{f_2(\rho)} - 1\right| \le c_2 \rho,$$

where the positive constant c_2 is independent of x.

The same inequality (with $|\rho|$ on the right-hand side) is also obtained for $-c_1^{-1} < \rho < 0$.

17

1.15. Theorem. *With the notation in Theorem 1.14,*

$$\frac{|x-y|}{\rho} \to 1 \quad \text{as} \quad y \to x, \quad \text{uniformly on } \partial S.$$

Proof. As above, we find that

$$\left|\frac{f_1(\rho)}{\rho} - 1\right| = \frac{\frac{1}{4}|\rho|^2 f''^2(\rho_1)}{\left[1 + \frac{1}{4}\rho^2 f''^2(\rho_1)\right]^{1/2} + 1} \leq c|\rho|^2,$$

where ρ_1 lies between 0 and ρ and c is independent of x.

1.16. Remark. If f is continuously differentiable in S_0, then we can write

$$\operatorname{grad} f(x) = \left[\tau(x)\frac{\partial}{\partial s(x)} + \nu(x)\frac{\partial}{\partial \nu(x)}\right] f(x), \quad x \in \partial S, \qquad (1.21)$$

where the notation $\partial/\partial s$ is preferred to $\partial/\partial \tau$.

1.4. Integrals with singular kernels

Let $C(S)$ and $C^1(S)$ be respectively the spaces of (real) continuous and continuously differentiable functions in S. We consider the set of all functions in $C(S)$ ($C^1(S)$) that are continuously extendable (continuously extendable together with their first order derivatives) to $\bar{S} = S \cup \partial S$, and denote by $C(\bar{S})$ ($C^1(\bar{S})$) the space of the corresponding extensions. The following assertion shows that this notation is justified.

1.17. Theorem. *Let $f \in C^1(S)$, and suppose that $f(x) \to l(\xi)$ and $\operatorname{grad} f(x) \to \lambda(\xi)$ as $S \ni x \to \xi \in \partial S$, where l and λ are continuous on ∂S. Then the function*

$$\tilde{f}(x) = \begin{cases} f(x), & x \in S, \\ l(x), & x \in \partial S, \end{cases}$$

has (one-sided) derivatives at all $x \in \partial S$ and

$$\operatorname{grad} \tilde{f}(x) = \begin{cases} \operatorname{grad} f(x), & x \in S, \\ \lambda(x), & x \in \partial S \end{cases}$$

(in other words, the operations of differentiation and extension to $\bar{S}$ commute for f).

Proof. Clearly, $\tilde{f} \in C(\bar{S}) \cap C^1(S)$. Consequently, for $\xi = (\xi_1, \xi_2) \in \partial S$ and $x = (x_1, \xi_2) \in S$, $x_1 \neq \xi_1$, in a sufficiently small neighbourhood of ξ we have

$$\left| \frac{\tilde{f}(x) - \tilde{f}(\xi)}{x_1 - \xi_1} - \lambda_1(\xi) \right| = \left| \frac{\partial}{\partial x_1} \tilde{f}(\eta) - \lambda_1(\xi) \right| = \left| \frac{\partial}{\partial x_1} f(\eta) - \lambda_1(\xi) \right|,$$

where $\eta = (\eta_1, \xi_2)$ with η_1 between x_1 and ξ_1. The result for $\partial \tilde{f}/\partial x_1$ now follows from the fact that the right-hand side tends to zero as $x \to \xi$. The argument for $\partial \tilde{f}/\partial x_2$ is similar.

1.18. Remark. The above spaces are also introduced for functions defined on ∂S. Let $f(x)$ be such a function, and let $x = \psi(s)$ in terms of the arc coordinate. Then for simplicity we also write $f(s) \equiv f(\psi(s))$. In this case the derivative of f is defined to be

$$f'(s) = f'(x) = \frac{d}{ds} f(x) = \lim_{t \to s} \frac{f(y) - f(x)}{t - s} = \lim_{t \to s} \frac{f(t) - f(s)}{t - s},$$

where $x, y \in \partial S$ and $y = \psi(t)$, provided that the limit exists. We specify that in what follows the notation f' for the derivative does not extend to position vectors. Thus, x' will denote a point on ∂S and not dx/ds.

Clearly, if f is defined and differentiable on a domain that includes ∂S, then the derivative along ∂S of the restriction of f to ∂S coincides with $\langle \operatorname{grad} f(x), \tau(x) \rangle$.

1.19. Definition. A function f defined on $\bar{S}$ is said to be *Hölder continuous* (with index $\alpha \in (0,1]$) on $\bar{S}$ if

$$|f(x) - f(y)| \leq c|x-y|^\alpha \quad \text{for all} \quad x, y \in \bar{S}, \tag{1.22}$$

where $c = \text{const} > 0$ is independent of x and y. If S is unbounded, then the above definition must hold on every bounded subdomain of S.

We denote by $C^{0,\alpha}(\bar{S})$ the vector space of (real) Hölder continuous (with index $\alpha \in (0,1]$) functions on $\bar{S}$, and by $C^{1,\alpha}(\bar{S})$ the subspace of $C^1(\bar{S})$ of functions whose first order derivatives belong to $C^{0,\alpha}(\bar{S})$.

1.20. Lemma. *If $0 < \beta < \alpha \leq 1$, then*
 (i) $C^{0,\alpha}(\bar{S}) \subset C^{0,\beta}(\bar{S})$;
 (ii) $fg \in C^{0,\beta}(\bar{S})$ *for all* $f \in C^{0,\alpha}(\bar{S})$ *and* $g \in C^{0,\beta}(\bar{S})$.

The proof consists in the verification of (1.22).

The spaces $C^{0,\alpha}(\partial S)$ and $C^{1,\alpha}(\partial S)$ are introduced similarly, with (1.22) required to hold for all $x, y \in \partial S$. In view of Lemma 1.3, we will not distinguish between $C^{0,\alpha}(\partial S)$ and $C^{0,\alpha}[0, |\partial S|]$, which is defined by means of the inequality

$$|f(s) - f(t)| \leq c|s-t|^\alpha \quad \text{for all} \quad s, t \in [0, |\partial S|].$$

Obviously, Lemma 1.20 also holds for functions on ∂S.

1.21. Remark. If f is bounded in $\bar{S}$, that is, $|f(x)| \leq M = \text{const}$ for all $x \in \bar{S}$, and (1.22) holds for all $x, y \in \bar{S}$ such that $|x - y| \leq \delta$, where $\delta = \text{const} > 0$, then it holds (possibly with a different c) for all $x, y \in \bar{S}$. This is easily seen, since for $|x - y| > \delta$ we can write

$$|f(x) - f(y)| \leq 2M < \frac{2M}{\delta^\alpha}|x-y|^\alpha.$$

1.22. Remark. If $\varphi \in C^{0,\alpha}(\partial S)$ as a function of $x = \psi(s)$, then, by Lemma 1.3, $\varphi \in C^{0,\alpha}(\partial S)$ also as a function of s, and vice versa.

1.23. Definition. A two-point function $k(x, y)$ defined and continuous for all $x \in S_0$ ($x \in \partial S$) and $y \in \partial S$, $x \neq y$, is called a γ-*singular kernel in* S_0 (*on* ∂S), $\gamma \in [0, 1]$, if there is a $p = \text{const} > 0$, which may depend on ∂S, such that
$$|k(x, y)| \leq p|x - y|^{-\gamma}$$
for all $x \in S_0$ ($x \in \partial S$), $x \neq y$.

If, in addition,
$$|k(x, y) - k(x', y)| \leq p|x - x'| \, |x - y|^{-\gamma - 1}$$
for all $x, x' \in S_0$ ($x, x' \in \partial S$) and $y \in \partial S$ satisfying $0 < |x - x'| < \frac{1}{2}|x - y|$, then $k(x, y)$ is called a *proper* γ-*singular kernel in* S_0 (*on* ∂S).

We extend this definition to two-point matrix functions by requiring each component to satisfy the necessary properties.

1.24. Remark. A kernel may have a lower 'singularity index' γ when it is considered on ∂S rather than in S_0. For example, the function $k(x, y) = \partial \ln |x - y|/\partial \nu(y)$ is a proper 1-singular kernel in S_0, but, by Lemma 1.1, a proper 0-singular kernel on ∂S.

1.25. Lemma. *If $k(x, y)$ is γ-singular in S_0, $\gamma \in [0, 1]$, and continuously differentiable with respect to x_α for all $x \in S_0$ and $y \in \partial S$, $x \neq y$, and if the kernels $|x - y|[\partial k(x, y)/\partial x_\alpha]$ are γ-singular in S_0, then $k(x, y)$ is a proper γ-singular kernel in S_0.*

Proof. Let $x, x' \in S_0$ and $y \in \partial S$ be such that $0 < |x - x'| < \frac{1}{2}|x - y|$. For any x'' on the line between x and x' we have
$$|x'' - y| \geq |x - y| - |x - x''| > |x - y| - \frac{1}{2}|x - y| = \frac{1}{2}|x - y|,$$

consequently,

$$|k(x, y) - k(x', y)| \leq |x_\alpha - x'_\alpha| \left|\frac{\partial}{\partial x_\alpha} k(x'', y)\right| \leq p'|x - x'||x - y|^{-\gamma-1},$$

where $p' = \text{const}$ depends only on γ.

1.26. Remark. If $k(x, y)$ is a γ-singular kernel on ∂S, $\gamma \in [0, 1]$, and continuously differentiable with respect to the arc coordinate $s(x)$ of x at all points $x, y \in \partial S$, $x \neq y$, and if $|x - y|[\partial k(x, y)/\partial s(x)]$ is γ-singular on ∂S, then $k(x, y)$ is a proper γ-singular kernel on ∂S. The proof of this statement is similar to that of Lemma 1.25, use also being made of Remark 1.5.

The following assertion is proved by direct verification of the required properties.

1.27. Lemma. (i) *If $k_1(x, y)$ is 0-singular and $k_2(x, y)$ is γ-singular, $\gamma \in [0, 1]$, then $k_1(x, y)k_2(x, y)$ is γ-singular.*

(ii) *If $k_1(x, y)$ is γ_1-singular and $k_2(x, y)$ is γ_2-singular, $0 \leq \gamma_1 \leq \gamma_2 \leq 1$, then $k_1(x, y) + k_2(x, y)$ is γ_2-singular.*

1.28. Remark. Lemma 1.27 also holds with 'singular' replaced by 'proper singular' in its statement.

1.29. Theorem. *If $k(x, y)$ is a γ-singular kernel on ∂S, $\gamma \in [0, 1)$, then the function*

$$f(x) = \int_{\partial S} k(x, y)\, ds(y) \tag{1.23}$$

is continuous on ∂S.

Proof. Let $x, a, b, y \in \partial S$, $x = \psi(s)$, $a = \psi(s - \varepsilon_1)$, $b = \psi(s + \varepsilon_2)$, $y = \psi(t)$, where $\varepsilon_1, \varepsilon_2 > 0$ are arbitrarily small, and let

$$I_\varepsilon(t) = \int_b^a |s - t|^{-\gamma} \, dt,$$

$$I(t) = \int_{\partial S} |s - t|^{-\gamma} \, dt.$$

Clearly,

$$|I(t) - I_\varepsilon(t)| = \frac{1}{1 - \gamma}(\varepsilon_1^{1-\gamma} + \varepsilon_2^{1-\gamma}),$$

therefore, $I_\varepsilon(t) \to I(t)$ uniformly with respect to t as $\varepsilon \to 0$. Since, by Definition 1.23 and Lemma 1.3,

$$|k(x, y)| \leq c|x - y|^{-\gamma} \leq c|s - t|^{-\gamma}$$

for all $x, y \in \partial S$, $x \neq y$, the improper integral (1.23) converges uniformly with respect to $x \in \partial S$, and the assertion follows from a well-known theorem of analysis [38].

1.30. Theorem. *If $k(x, y)$ is a proper γ-singular kernel in S_0 (on ∂S), $\gamma \in [0, 1)$, and $\varphi \in C(\partial S)$, then the function*

$$K(x) = \int_{\partial S} k(x, y)\varphi(y) \, ds(y), \quad x \in S_0 \ (x \in \partial S),$$

belongs to $C^{0,\beta}(S_0)$, with $\beta = 1 - \gamma$ for $\gamma \in (0, 1)$ and any $\beta \in (0, 1)$ for $\gamma = 0$. In addition,

$$\sup_{\substack{x,x' \in S_0 \ (\partial S) \\ x \neq x'}} \frac{|K(x) - K(x')|}{|x - x'|^\beta} \leq c \sup_{x \in \partial S} |\varphi(x)|,$$

where $c = \text{const} > 0$ may depend on γ.

23

Proof. $K(x)$ is obviously an improper integral for $x \in \partial S$.

Let $\Sigma_{x,r}$, Σ_1 and Σ_2 be the sets defined by (1.12), (1.15) and (1.16). In view of Remark 1.21, we may consider x, $x' \in S_0$ satisfying (1.18).

Setting $x = \xi + \sigma \nu(\xi)$ and $x' = \xi' + \sigma' \nu(\xi')$, $\xi, \xi' \in \partial S$, we can write

$$K(x) - K(x') = I_1 + I_2 + I_3,$$

where, by Definition 1.19, Remark 1.12, and Lemmas 1.8–1.11,

$$|I_1| = \left| \int_{\Sigma_1} [k(x, y) - k(x', y)] \varphi(y) \, ds(y) \right|$$

$$\leq c_1 \sup_{x \in \partial S} |\varphi(x)| \int_{\Sigma_1} (|x - y|^{-\gamma} + |x' - y|^{-\gamma}) \, ds(y)$$

$$\leq c_2 |x - x'|^{1-\gamma} \sup_{x \in \partial S} |\varphi(x)|,$$

$$|I_2| = \left| \int_{\Sigma_2} [k(x, y) - k(x', y)] \varphi(y) \, ds(y) \right|$$

$$\leq c_3 |x - x'| \sup_{x \in \partial S} |\varphi(x)| \int_{\Sigma_2} |x - y|^{-\gamma-1} \, ds(y)$$

$$\leq c_4 |x - x'|^{1-\gamma} \sup_{x \in \partial S} |\varphi(x)| \quad \text{if} \quad \gamma \in (0, 1),$$

$$|I_2| \leq c_5 |x - x'| |\ln |x - x'|| \sup_{x \in \partial S} |\varphi(x)| \quad \text{if} \quad \gamma = 0,$$

$$|I_3| = \left| \int_{\partial S \setminus \Sigma_{\xi,r}} [k(x, y) - k(x', y)] \varphi(y) \, ds(y) \right|$$

$$\leq c_6 |x - x'| \sup_{x \in \partial S} |\varphi(x)| \int_{\partial S \setminus \Sigma_{\xi,r}} |x - y|^{-\gamma-1} \, ds(y)$$

$$\leq c_7 r^{-\gamma-1} |\partial S| \, |x - x'| \sup_{x \in \partial S} |\varphi(x)|$$

$$= c_8 |x - x'| \sup_{x \in \partial S} |\varphi(x)|.$$

The assertion now follows from the fact that the constants $c_1, \ldots, c_8 > 0$ are independent of x and x' (although they may depend on γ).

The result is established for $x, x' \in \partial S$ as a particular case of the above, by setting $x = \xi$ and $x' = \xi'$.

1.31. Remark. It is obvious that Theorem 1.30 holds if the kernel $k(x, y)$ is continuous on $S_0 \times \partial S$ ($\partial S \times \partial S$).

1.32. Theorem. *If $k(x, y)$ is a proper 1-singular kernel in S_0 (on ∂S), $\varphi \in C^{0,\alpha}(\partial S)$, $\alpha \in (0, 1]$, and*

$$\Phi(x) = \int_{\partial S} k(x, y)[\varphi(y) - \varphi(\xi)] \, ds(y), \tag{1.24}$$

where $x = \xi + \sigma\nu(\xi) \in S_0$ ($x = \xi \in \partial S$), then $\Phi \in C^{0,\beta}(S_0)$ ($\Phi \in C^{0,\beta}(\partial S)$) for any $\beta \in (0, \alpha)$. If, in addition, $\alpha \in (0, 1)$ and

$$\left| \int_{\partial S \setminus \Sigma_{\xi,\delta}} k(x, y) \, ds(y) \right| \leq c = \text{const} > 0 \tag{1.25}$$

for all $x \in S_0$ ($x \in \partial S$) and all $0 < \delta < r$, then $\Phi \in C^{0,\alpha}(S_0)$ ($\Phi \in C^{0,\alpha}(\partial S)$).

Proof. Clearly, Φ exists as an improper integral if $x \in \partial S$, and, by Theorem 1.29, is continuous on ∂S.

As in the proof of Theorem 1.30, let $x, x' \in S_0$ be chosen so that (1.18) holds. Writing

$$\Phi(x) - \Phi(x')$$
$$= \int_{\Sigma_1} \{k(x, y)[\varphi(y) - \varphi(\xi)] - k(x', y)[\varphi(y) - \varphi(\xi')]\} \, ds(y)$$

$$+ \int_{\Sigma_2} \{[k(x, y) - k(x', y)] [\varphi(y) - \varphi(\xi')]$$
$$- k(x, y)[\varphi(\xi) - \varphi(\xi')]\} ds(y)$$
$$+ \int_{\partial S \setminus \Sigma_{\xi,r}} \{[k(x, y) - k(x', y)] [\varphi(y) - \varphi(\xi')]$$
$$- k(x, y)[\varphi(\xi) - \varphi(\xi')]\} ds(y)$$
$$= I_1 + I_2 + I_3,$$

from Definition 1.19, Remark 1.12 and Lemma 1.11 we now find that

$$|I_1| \le c_1 \int_{\Sigma_1} (|\xi - y|^{\alpha-1} + |\xi' - y|^{\alpha-1}) \, ds(y) \le c_2 |x - x'|^\alpha,$$

$$|I_2| \le c_3 |x - x'| \int_{\Sigma_2} |\xi - y|^{\alpha-2} \, ds(y) + c_4 |\xi - \xi'|^\alpha \int_{\Sigma_2} |\xi - y|^{-1} \, ds(y)$$
$$\le c_5 |x - x'|^\alpha + c_6 |x - x'|^\alpha |\ln |x - x'|| \quad \text{for } \alpha \in (0, 1),$$

$$|I_2| \le c_3 |x - x'| \int_{\Sigma_2} |\xi - y|^{-1} \, ds(y) + c_4 |\xi - \xi'|^\alpha \int_{\Sigma_2} |\xi - y|^{-1} \, ds(y)$$
$$\le c_7 |x - x'|^\alpha |\ln |x - x'|| \quad \text{for } \alpha = 1,$$

$$|I_3| \le c_8 |x - x'| \int_{\partial S \setminus \Sigma_{\xi,r}} |\xi - y|^{\alpha-2} \, ds(y) + c_9 |\xi - \xi'|^\alpha \int_{\partial S \setminus \Sigma_{\xi,r}} |\xi - y|^{-1} \, ds(y)$$
$$\le c_8 r^{\alpha-2} |\partial S| |x - x'| + c_{10} r^{-1} |\partial S| |x - x'|^\alpha \le c_{11} |x - x'|^\alpha,$$

where $c_1, \ldots, c_{11}$ are positive constants independent of x and x'.

This proves the first part of the assertion. For the second part we combine the last terms in I_2 and I_3 and use the fact that $\int_{\partial S \setminus \Sigma_1} k(x, y) \, ds(y)$ is bounded for all $x, x' \in S_0$ satisfying the conditions of the theorem. (See Remark 1.38 below for a full explanation of this detail.)

The result for $x, x' \in \partial S$ is again obtained by setting $x = \xi$ and $x' = \xi'$.

1.33. Remark. By Theorem 1.29, the estimate (1.25) holds on ∂S if $k(x, y)$ is a γ-singular kernel on ∂S, $\gamma \in [0, 1)$.

1.34. Theorem. *Let $k(x, y)$ be a β-singular kernel on ∂S, $\beta \in [0, 1)$, such that*

(i) $g(x) = \partial [\int_{\partial S} k(x, y) \, ds(y)] / \partial s$ *exists for all $x \in \partial S$ and $g \in C(\partial S)$;*

(ii) $[k(x', y) - k(x, y)](s' - s)^{-1} = k_0(x, y) + O(|s' - s| |x - y|^{-\gamma - 2})$ *for all $x, x', y \in \partial S$, $0 < |x - x'| < \frac{1}{2}|x - y|$, where $x = \psi(s)$, $x' = \psi(s')$, and $|x - y| k_0(x, y)$ is a γ-singular kernel on ∂S, $\gamma \in [0, 1)$.*

If $\varphi \in C^{0,\alpha}(\partial S)$, $\alpha \in (\beta, 1]$, $\alpha > \gamma$, then the function

$$F(x) = \int_{\partial S} k(x, y) \varphi(y) \, ds(y), \quad x \in \partial S,$$

belongs to $C^1(\partial S)$ and

$$\frac{\partial}{\partial s} F(x) = \int_{\partial S} k_0(x, y) [\varphi(y) - \varphi(x)] \, ds(y) + \varphi(x) g(x). \quad (1.26)$$

Proof. Let $G(x)$ be the function on the right-hand side in (1.26). By Theorem 1.29, $F(x)$ and the first term in $G(x)$ exist as improper integrals and are continuous on ∂S; the second term in $G(x)$ is continuous by assumption.

Let $x, x' \in \partial S$ be such that $0 < |x - x'| < \frac{1}{8}r$, with r satisfying (1.18). We have

$$[F(x') - F(x)](s' - s)^{-1} - G(x)$$
$$= (s' - s)^{-1} \int_{\Sigma_1} \{k(x', y)[\varphi(y) - \varphi(x')]$$
$$- k(x, y)[\varphi(y) - \varphi(x)]\} \, ds(y)$$

27

$$+ (s'-s)^{-1}[\varphi(x')-\varphi(x)]\int_{\Sigma_1} k(x',y)\,ds(y)$$

$$-\int_{\Sigma_1} k_0(x,y)[\varphi(y)-\varphi(x)]\,ds(y)$$

$$+\int_{\Sigma_2}\{[k(x',y)-k(x,y)](s'-s)^{-1}$$
$$\qquad\qquad\qquad -k_0(x,y)\}[\varphi(y)-\varphi(x)]\,ds(y)$$

$$+\int_{\partial S\setminus\Sigma_{x,r}}\{[k(x',y)-k(x,y)](s'-s)^{-1}$$
$$\qquad\qquad\qquad -k_0(x,y)\}[\varphi(y)-\varphi(x)]\,ds(y)$$

$$+\varphi(x)\left\{\left[\int_{\partial S} k(x',y)\,ds(y)-\int_{\partial S} k(x,y)\,ds(y)\right](s'-s)^{-1}-g(x)\right\}$$

$$= I_1 + I_2 + I_3 + I_4 + I_5 + I_6.$$

By Definition 1.19, Remark 1.12 and Lemmas 1.8 and 1.11,

$$|I_1| \le c_1|s'-s|^{-1}\int_{\Sigma_1}(|x'-y|^{\alpha-\beta}+|x-y|^{\alpha-\beta})\,ds(y) \le c_2|s'-s|^{\alpha-\beta},$$

$$|I_2| \le c_3|s'-s|^{\alpha-1}\int_{\Sigma_1}|x'-y|^{-\beta}\,ds(y) \le c_4|s'-s|^{\alpha-\beta},$$

$$|I_3| \le c_5\int_{\Sigma_1}|x-y|^{\alpha-\gamma-1}\,ds(y) \le c_6|s'-s|^{\alpha-\gamma}.$$

By Lemma 1.9, $y \in \Sigma_2$ implies that $|x-x'| < \frac{1}{2}|x-y|$. Hence,

$$|I_4| \le c_7|s'-s|\int_{\Sigma_2}|x-y|^{\alpha-\gamma-2}\,ds(y) \le c_8|s'-s|^{\alpha-\gamma}.$$

Finally, by Lemma 1.10,

$$|I_5| \le c_9|s'-s|\int_{\partial S\setminus\Sigma_{x,r}}|x-y|^{\alpha-\gamma-2}\,ds(y) \le c_{10}|s'-s|.$$

Since all the constants $c_1, \ldots, c_{10} > 0$ are independent of x and x', we find that $I_j \to 0$ as $s' - s \to 0$, $j = 1, \ldots, 5$.

In addition, by our assumption (i), $I_6 \to 0$ as $s' - s \to 0$, which proves that $F'(x)$ exists for all $x \in \partial S$ and is given by (1.26), whose right-hand side is obviously a continuous function on ∂S.

1.35. Remark. If $g \in C^{0,\alpha}(\partial S)$ and $k_0(x,y)$ is a proper 1-singular kernel on ∂S, then, by Theorem 1.32, $F \in C^{1,\beta}(\partial S)$ for any $\beta \in (0,\alpha)$. If, furthermore, $\alpha \in (0,1)$ and $k_0(x,y)$ satisfies the estimate (1.25), then $F \in C^{1,\alpha}(\partial S)$.

1.36. Remark. In practice it is helpful to have some easily checked condition in place of the assumption (ii) in Theorem 1.34. Suppose that $k(x,y)$ is continuously differentiable with respect to $s(x)$ for all $x, y \in \partial S$, $x \neq y$, and that $|x - y|[\partial k(x,y)/\partial s(x)]$ is a proper γ-singular kernel on ∂S, $\gamma \in [0,1)$. Then for $x, x', y \in \partial S$ such that $0 < |x - x'| < \frac{1}{2}|x - y|$ we have

$$[k(x', y) - k(x, y)](s' - s)^{-1}$$
$$= \frac{\partial}{\partial s(x)} k(x, y) + \left[\frac{\partial}{\partial s(x)} k(x'', y) - \frac{\partial}{\partial s(x)} k(x, y) \right],$$

where $x'' \in \partial S$ lies between x and x'. Since

$$\left| \frac{\partial}{\partial s(x)} k(x'', y) - \frac{\partial}{\partial s(x)} k(x, y) \right| \leq c|s - s'| |x - y|^{-\gamma - 2},$$

it follows that under the above conditions the assumption (ii) in Theorem 1.34 holds with $k_0(x, y) = \partial k(x, y)/\partial s(x)$.

1.37. Definition. Let $k(x,y)$ be defined and continuous for all points $x, y \in \partial S$, $x \neq y$. We say that $\int_{\partial S} f(x, y)\, ds(y)$ exists as *principal value* if

$$\lim_{\delta \to 0} \int_{\partial S \setminus \Sigma_{x,\delta}} k(x, y) \, ds(y) \qquad (1.27)$$

exists for all $x \in \partial S$.

Obviously, an ordinary (even improper) integral exists as principal value, but the converse is not true in general.

In what follows the principal value of an integral (if it exists) is denoted by the same symbol as an ordinary integral, the difference in meaning being either explicitly stated, or understood from the context as the only possible alternative.

1.38. Remark. Let $k(x, y)$ be a 1-singular kernel on ∂S, and let a_1, a_2 and b_1, b_2, $a_1 \le b_1 < b_2 \le a_2$, be the arc coordinates of the end-points of the sets $\Sigma_{x,\delta}$ and

$$\Gamma_{x,\delta} = \{y \in \partial S : |t - s| \le \delta\}, \qquad (1.28)$$

respectively, where $x = \psi(s)$ and $y = \psi(t)$. Since

$$\left| \int_{\partial S \setminus \Gamma_{x,\delta}} k(x, y) \, ds(y) - \int_{\partial S \setminus \Sigma_{x,\delta}} k(x, y) \, ds(y) \right|$$

$$= \left| \int_{\Sigma_{x,\delta} \setminus \Gamma_{x,\delta}} k(x, y) \, ds(y) \right| \le c \left(\int_{a_1}^{b_1} |s - t|^{-1} \, dt + \int_{b_2}^{a_2} |s - t|^{-1} \, dt \right)$$

$$= c \ln \left(\frac{s - a_1}{s - b_1} \cdot \frac{a_2 - s}{b_2 - s} \right) = c \ln \left(\frac{s - a_1}{\delta} \cdot \frac{a_2 - s}{\delta} \right),$$

Theorem 1.14 implies that if $\int_{\partial S} k(x, y) \, ds(y)$ exists in the sense of principal value, then its definition can equivalently be given as

$$\lim_{\delta \to 0} \int_{\partial S \setminus \Gamma_{x,\delta}} k(x, y) \, ds(y).$$

Moreover, if the limit (1.27) exists uniformly for all $x \in \partial S$, then so does the above one, and vice versa.

1.39. Remark. Let ρ be the local coordinate of $y \in \Sigma_{x,r}$ measured from x along the support line of $\tau(x)$ (see Remark 1.13), and consider the set
$$\Lambda_{x,\delta} = \{y \in \Sigma_{x,r} : |\rho| \leq \delta\}, \quad \delta < \tfrac{1}{2}r.$$

Since $\delta < \tfrac{1}{2}r$, all points in the neighbourhood of x such that $|\rho| \leq \delta$ belong to $\Sigma_{x,r}$. Denoting by $-a$ and b, $a, b > 0$, the ρ-coordinates of the end-points of $\Sigma_{x,\delta}$, we find that for a 1-singular kernel $k(x,y)$ on ∂S

$$\left| \int_{\partial S \setminus \Sigma_{x,\delta}} k(x,y)\, ds(y) - \int_{\partial S \setminus \Lambda_{x,\delta}} k(x,y)\, ds(y) \right|$$

$$= \left| \int_{\Lambda_{x,\delta} \setminus \Sigma_{x,\delta}} k(x,y)\, ds(y) \right| \leq c_1 \int_{\Lambda_{x,\delta} \setminus \Sigma_{x,\delta}} |x-y|^{-1}\, ds(y)$$

$$\leq c_2 \left(\int_{-\delta}^{-a} (-\rho)^{-1}\, d\rho + \int_{b}^{\delta} \rho^{-1}\, d\rho \right) = c_2 \ln\left(\frac{\delta}{a} \cdot \frac{\delta}{b} \right),$$

where c_2 does not depend on x. Consequently, by Theorem 1.15, if $\int_{\partial S} k(x,y)\, ds(y)$ exists in the sense of principal value, then it can also be defined as
$$\lim_{\delta \to 0} \int_{\partial S \setminus \Lambda_{x,\delta}} k(x,y)\, ds(y).$$

Furthermore, from Theorem 1.15 it follows that the existence of either of these two equivalent limits uniformly with respect to $x \in \partial S$ implies the same property for the other one.

1.40. Theorem. *Let $k(x, y)$ be a proper 1-singular kernel in S_0 that is γ-singular on ∂S, $\gamma \in [0, 1)$, and let*

$$f(x) = \int_{\partial S} k(x, y) \, ds(y), \quad x \in S_0 \setminus \partial S,$$
$$f_0(x) = \int_{\partial S} k(x, y) \, ds(y), \quad x \in \partial S,$$ (1.29)

and

$$F(x) = \int_{\partial S} k(x, y) \varphi(y) \, ds(y), \quad x \in S_0 \setminus \partial S,$$
$$F_0(x) = \int_{\partial S} k(x, y) \varphi(y) \, ds(y), \quad x \in \partial S,$$ (1.30)

where $\varphi \in C^{0,\alpha}(\partial S)$, $\alpha \in (0, 1]$. Also, consider the functions

$$f^+(x) = \begin{cases} f(x), & x \in S_0^+, \\ l(x) + f_0(x), & x \in \partial S, \end{cases}$$
$$f^-(x) = \begin{cases} f(x), & x \in S_0^-, \\ -l(x) + f_0(x), & x \in \partial S, \end{cases}$$ (1.31)

and

$$F^+(x) = \begin{cases} F(x), & x \in S_0^+, \\ l(x)\varphi(x) + F_0(x), & x \in \partial S, \end{cases}$$
$$F^-(x) = \begin{cases} F(x), & x \in S_0^-, \\ -l(x)\varphi(x) + F_0(x), & x \in \partial S, \end{cases}$$ (1.32)

where $l \in C^{0,\alpha}(\partial S)$. If $f^+ \in C^{0,\alpha}(\bar{S}_0^+)$ and $f^- \in C^{0,\alpha}(\bar{S}_0^-)$, then $F^+ \in C^{0,\beta}(\bar{S}_0^+)$ and $F^- \in C^{0,\beta}(\bar{S}_0^-)$, with $\beta = \alpha$ for $\alpha \in (0, 1)$ and any $\beta \in (0, 1)$ for $\alpha = 1$.

Proof. From the properties of $k(x, y)$ it is clear that f_0 and F_0 are improper integrals.

To prove the statement for F^+ it suffices to consider $x, x' \in \bar{S}_0^+$ satisfying (1.18). Let $x = \xi + \sigma \nu(\xi) \in S_0^+$, $\xi \in \partial S$, and $x' = \xi' \in \partial S$. Then

$$\int_{\partial S} k(x, y)\varphi(y) \, ds(y) - l(x')\varphi(x') - \int_{\partial S} k(x', y)\varphi(y) \, ds(y)$$

$$= \int_{\partial S} k(x, y)[\varphi(y) - \varphi(\xi)] \, ds(y) - \int_{\partial S} k(x', y)[\varphi(y) - \varphi(x')] \, ds(y)$$

$$+ [\varphi(\xi) - \varphi(x')] \int_{\partial S} k(x, y) \, ds(y)$$

$$+ \varphi(x') \left[\int_{\partial S} k(x, y) \, ds(y) - l(x') - \int_{\partial S} k(x', y) \, ds(y) \right], \quad (1.33)$$

that is,

$$F^+(x) - F^+(x') = \Phi(x) - \Phi(x') + [\varphi(\xi) - \varphi(\xi')] f^+(x)$$
$$+ [f^+(x) - f^+(x')]\varphi(\xi'), \quad (1.34)$$

where Φ is given by (1.24). The equality (1.34) is similarly obtained when $x, x' \in S_0^+$, $x, x' \in \partial S$, or $x \in \partial S$, $x' \in S_0^+$. Since, by our assumption, both f_0 and f are bounded, (1.29) shows that $k(x, y)$ satisfies the estimate (1.25). The assertion now follows from (1.34) and Theorem 1.32.

F^- is treated analogously.

This theorem can be generalized to certain 1-singular kernels on ∂S.

1.41. Definition. A 1-singular kernel on ∂S is called *integrable* if $\int_{\partial S} k(x, y) \, ds(y)$ exists as principal value for all $x \in \partial S$, and *uniformly integrable* if the integral in (1.27) converges uniformly with respect to $x \in \partial S$.

1.42. Remark. If the kernel $k(x, y)$ is uniformly integrable, then $\int_{\partial S} k(x, y) \, ds(y)$ is continuous on ∂S. This is shown by writing the principal value of the integral as the sum of a uniformly convergent infinite series.

Also, it is clear that a γ-singular kernel on ∂S, $\gamma \in [0, 1)$, is uniformly integrable, and that any uniformly integrable kernel satisfies (1.25) on ∂S.

1.43. Theorem. *If $k(x, y)$ is 1-singular on ∂S and integrable, and i $\varphi \in C^{0,\alpha}(\partial S)$, $\alpha \in (0, 1]$, then the integral*

$$\int_{\partial S} k(x, y) \varphi(y) \, ds(y)$$

exists in the sense of principal value for all $x \in \partial S$. If $k(x, y)$ is uniformly integrable, then the above principal value exists uniformly with respect to $x \in \partial S$.

Proof. We write

$$\int_{\partial S \setminus \Sigma_{x,\delta}} k(x, y) \varphi(y) \, ds(y) = \int_{\partial S \setminus \Sigma_{x,\delta}} k(x, y) [\varphi(y) - \varphi(x)] \, ds(y)$$

$$+ \varphi(x) \int_{\partial S \setminus \Sigma_{x,\delta}} k(x, y) \, ds(y).$$

The result follows from the fact that, as $\delta \to 0$, the first term on the right-hand side converges uniformly since its integrand is $O(|x - y|^{\alpha-1})$.

1.44. Theorem. *Suppose that*
  (i) *$k(x, y)$ is a proper 1-singular kernel in S_0 that is integable on ∂S;*
  (ii) *f^+ and f^- defined by (1.31), where $l \in C^{0,\alpha}(\partial S)$, $\alpha \in (0, 1]$, and f_0 is understood as principal value, belong respectively to $C^{0,\alpha}(\bar{S}_0^+)$ and $C^{0,\alpha}(\bar{S}_0^-)$.*

Then the functions F^+ and F^- defined by (1.32), where $\varphi \in C^{0,\alpha}(\partial S)$ and F_0 is understood as principal value, belong to $C^{0,\beta}(\bar{S}_0^+)$ and $C^{0,\beta}(\bar{S}_0^-)$, respectively, with $\beta = \alpha$ for $\alpha \in (0,1)$ and any $\beta \in (0,1)$ for $\alpha = 1$.

Proof. By Theorem 1.43, F_0 exists in the sense of principal value for all $x \in \partial S$.

As in the proof of Theorem 1.40, let $x, x' \in \bar{S}_0^+$, $x \neq x'$. If $x, x' \in S_0^+$, the equality (1.34) is established immediately. If $x \in S_0^+$, $x' \in \partial S$ (or $x \in \partial S$, $x' \in S_0^+$), we write (1.33) with the integrals extended over $\partial S \setminus \Sigma_{x',\delta}$ ($\partial S \setminus \Sigma_{x,\delta}$) in the first instance, then let $\delta \to 0$. Noting that the limit of the second term on the right-hand side coincides with the improper integral $\Phi(x')$ ($\Phi(x)$), we again arrive at (1.34). Finally, we see that this is also true if both $x, x' \in \partial S$, when the integrals in (1.33) are initially extended over $\partial S \setminus (\Sigma_{x,\delta} \cup \Sigma_{x',\delta})$. Hence, (1.34) holds for all $x, x' \in \bar{S}_0^+$, $x \neq x'$, and the result follows from the assumptions (i) and (ii) and Theorem 1.32.

The reasoning is similar in the case of $\bar{S}_0^-$.

1.5. The harmonic potentials

In what follows we examine the Hölder continuity and continuous differentiability on $\bar{S}^+$ and $\bar{S}^-$ of functions that are analytic in S^+ and S^-. Hence, it suffices to consider the behaviour of such functions in the boundary layer S_0.

We begin by giving a brief account of the main properties of the harmonic potentials, which will be required at a later stage in the proceedings.

The *single layer potential* is defined by

$$(v(\varphi))(x) = -\int_{\partial S} (\ln|x-y|)\varphi(y)\,ds(y), \qquad (1.35)$$

and the *double layer potential* by

$$(w(\varphi))(x) = -\int_{\partial S} \left[\frac{\partial}{\partial \nu(y)} \ln|x-y|\right] \varphi(y)\, ds(y), \qquad (1.36)$$

where φ is the density. The explicit mention of the density is omitted when this does not create ambiguity.

We denote by S^+ the finite domain bounded by ∂S and set $S^- = \mathbf{R}^2 \setminus \bar{S}^+$.

1.45. Theorem. *If $\varphi \in C(\partial S)$, then $v(\varphi) \in C^{0,\alpha}(\mathbf{R}^2)$ for any index $\alpha \in (0,1)$.*

Proof. The assertion follows from Theorem 1.30 in view of the fact that, as can easily be verified by means of Lemma 1.25, the kernel $k(x,y) = -\ln|x-y|$ of v is a proper γ-singular kernel in S_0 for any $\gamma \in (0,1)$.

1.46. Theorem. *If $\varphi \in C^{0,\alpha}(\partial S)$, $\alpha \in (0,1]$, then the restrictions of $w(\varphi)$ to S^+ and S^- have $C^{0,\beta}$-extensions to $\bar{S}^+$ and $\bar{S}^-$, respectively, with $\beta = \alpha$ for $\alpha \in (0,1)$ and any $\beta \in (0,1)$ for $\alpha = 1$. These extensions are given by*

$$\begin{aligned} w^+(x) &= \begin{cases} w(x), & x \in S^+, \\ -\pi\varphi(x) + w_0(x), & x \in \partial S, \end{cases} \\ w^-(x) &= \begin{cases} w(x), & x \in S^-, \\ \pi\varphi(x) + w_0(x), & x \in \partial S, \end{cases} \end{aligned} \qquad (1.37)$$

where

$$w_0(x) = -\int_{\partial S} \left[\frac{\partial}{\partial \nu(y)} \ln|x-y|\right] \varphi(y)\, ds(y), \quad x \in \partial S. \qquad (1.38)$$

Proof. Applying Lemmas 1.25 and 1.1, we can easily verify that

$$k(x,y) = -\frac{\partial}{\partial \nu(y)} \ln|x-y| = \frac{\langle \nu(y), x-y \rangle}{|x-y|^2}$$

is a proper 1-singular kernel in S_0 and 0-singular on ∂S. Consequently, w_0 is an improper integral.

Let $x \in S^+$, and consider a disk $\sigma_{x,\delta} \subset S^+$ with centre at x and radius δ sufficiently small. Using the Divergence Theorem in $S^+ \setminus \sigma_{x,\delta}$ and the fact that $\ln|x-y|$ is a solution of the Laplace equation for $x \neq y$, we find that

$$0 = \int_{S^+ \setminus \sigma_{x,\delta}} \Delta(y) \ln|x-y|\, da(y)$$

$$= \int_{\partial S} \frac{\partial}{\partial \nu(y)} \ln|x-y|\, ds(y) - \int_{\partial \sigma_{x,\delta}} \frac{\partial}{\partial \nu(y)} \ln|x-y|\, ds(y)$$

$$= \int_{\partial S} \frac{\partial}{\partial \nu(y)} \ln|x-y|\, ds(y) - 2\pi,$$

where $\partial \sigma_{x,\delta}$ is the circular boundary of $\sigma_{x,\delta}$. Hence,

$$\int_{\partial S} \frac{\partial}{\partial \nu(y)} \ln|x-y|\, ds(y) = 2\pi, \quad x \in S^+. \tag{1.39}$$

The procedure is similar for $x \in \partial S$, except that in this case $\sigma_{x,\delta}$ is replaced by $\sigma_{x,\delta} \cap S^+$ and $\partial \sigma_{x,\delta}$ by its part lying in S^+. It is not difficult to show that for δ small the length of this part is equal to $\pi \delta + O(\delta^2)$, which leads to

$$\int_{\partial S} \frac{\partial}{\partial \nu(y)} \ln|x-y|\, ds(y) = \pi, \quad x \in \partial S. \tag{1.40}$$

Finally, the direct application of the Divergence Theorem yields

$$\int_{\partial S} \frac{\partial}{\partial \nu(y)} \ln|x-y|\, ds(y) = 0, \quad x \in S^-. \tag{1.41}$$

In view of these integrals and the expression of $k(x, y)$ we now see that

$$f(x) = \int_{\partial S} k(x, y) \, ds(y) = \begin{cases} -2\pi, & x \in S^+, \\ 0, & x \in S^-, \end{cases}$$

$$f_0(x) = \int_{\partial S} k(x, y) \, ds(y) = -\pi, \quad x \in \partial S. \tag{1.42}$$

From (1.31) with $l(x) = -\pi$, $x \in \partial S$, we obtain

$$f^+(x) = -2\pi, \quad x \in \bar{S}_0^+,$$
$$f^-(x) = 0, \quad x \in \bar{S}_0^-.$$

Since $f^+ \in C^{0,\alpha}(\bar{S}_0^+)$ and $f^- \in C^{0,\alpha}(\bar{S}_0^-)$, the result follows from Theorem 1.40.

1.47. Remark. Theorem 1.46 implies that if $\varphi \in C^{0,\alpha}(\partial S)$, then, as $S^\pm \ni x' \to x \in \partial S$, $w(\varphi)$ tends to finite limits given by

$$w^\pm(x) = \mp \pi \varphi(x) - \int_{\partial S} \left[\frac{\partial}{\partial \nu(y)} \ln |x - y| \right] \varphi(y) \, ds(y), \quad x \in \partial S, \tag{1.43}$$

where the last term is an improper integral. It can be shown [4] that $w(\varphi)$ can also be extended by continuity to $\bar{S}^+$ and $\bar{S}^-$ if $\varphi \in C(\partial S)$, but then the two extensions w^+ and w^- are merely continuous.

1.48. Theorem. *If $\varphi \in C^{0,\alpha}(\partial S)$, $\alpha \in (0, 1]$, then the first order derivatives of $v(\varphi)$ in S^+ and S^- have $C^{0,\beta}$-extensions to $\bar{S}^+$ and $\bar{S}^-$, respectively, with $\beta = \alpha$ for $\alpha \in (0, 1)$ and any $\beta \in (0, 1)$ for $\alpha = 1$. These extensions are given by*

$$(\operatorname{grad} v)^+(x) = \begin{cases} \operatorname{grad} v(x), & x \in S^+, \\ \pi \nu(x) \varphi(x) + (\operatorname{grad} v)_0(x), & x \in \partial S, \end{cases}$$

$$(\operatorname{grad} v)^-(x) = \begin{cases} \operatorname{grad} v(x), & x \in S^-, \\ -\pi \nu(x) \varphi(x) + (\operatorname{grad} v)_0(x), & x \in \partial S, \end{cases}$$

where

$$(\operatorname{grad} v)_0(x) = -\int_{\partial S} [\operatorname{grad}(x) \ln |x-y|] \varphi(y) \, ds(y), \quad x \in \partial S,$$

the integral being understood as principal value.

Proof. By checking the properties required in Lemma 1.25, we convince ourselves that
$$k(x, y) = -\operatorname{grad}(x) \ln |x-y|$$
is a proper 1-singular kernel in S_0 and on ∂S.

From (1.21) and the fact that
$$[\operatorname{grad}(x) + \operatorname{grad}(y)] \ln |x-y| = 0, \quad x \neq y,$$
it follows that
$$k(x, y) = \left[\frac{\partial}{\partial s(y)} \ln |x-y|\right] \tau(y) + \left[\frac{\partial}{\partial \nu(y)} \ln |x-y|\right] \nu(y), \quad x \neq y.$$

Consequently, using integration by parts and denoting by a and b the end-points of $\Sigma_{x,\delta}$, for $x \in \partial S$ we can write

$$\int_{\partial S \setminus \Sigma_{x,\delta}} k(x, y) \, ds(y) = \int_{\partial S \setminus \Sigma_{x,\delta}} \left[\frac{\partial}{\partial s(y)} \ln |x-y|\right] \tau(y) \, ds(y)$$
$$+ \int_{\partial S \setminus \Sigma_{x,\delta}} \left[\frac{\partial}{\partial \nu(y)} \ln |x-y|\right] \nu(y) \, ds(y)$$
$$= [\tau(a) - \tau(b)] \ln \delta - \int_{\partial S \setminus \Sigma_{x,\delta}} (\ln |x-y|) \kappa(y) \nu(y) \, ds(y)$$
$$+ \int_{\partial S \setminus \Sigma_{x,\delta}} \left[\frac{\partial}{\partial \nu(y)} \ln |x-y|\right] \nu(y) \, ds(y).$$

Since ∂S is a C^2-curve, the first term on the right-hand side tends to zero as $\delta \to 0$, while the other two tend to $(v(\kappa\nu))(x)$ and $-(w_0(\nu))(x)$, respectively. Therefore, $k(x, y)$ is integrable on ∂S and

$$f_0(x) = \int_{\partial S} k(x, y)\, ds(y) = (v(\kappa\nu))(x) - (w_0(\nu))(x), \quad x \in \partial S,$$

where f_0 is understood as principal value.

On the other hand, if $x \in S_0 \setminus \partial S$, then, again integrating by parts and taking (1.8) into account, we find that

$$f(x) = \int_{\partial S} k(x, y)\, ds(y) = (v(\kappa\nu))(x) - (w(\nu))(x), \quad x \in S_0 \setminus \partial S.$$

By Theorems 1.45 and 1.46, the function f is $C^{0,\alpha}$-extendable to $\bar{S}_0^+$ and $\bar{S}_0^-$ and the values of the corresponding extensions on ∂S are given by the formula

$$f^\pm(x) = (v(\kappa\nu))(x) \pm \pi\nu(x) + (w_0(\nu))(x) = \pm\pi\nu(x) + f_0(x), \quad x \in \partial S,$$

in other words, the expressions (1.31) with $l = \pi\nu \in C^{0,1}(\partial S)$. As stated, $f^+ \in C^{0,\alpha}(\bar{S}_0^+)$ and $f^- \in C^{0,\alpha}(\bar{S}_0^-)$. The assertion now follows from Theorem 1.44 with F and F_0 in (1.30) defined by

$$F(x) = -\int_{\partial S} [\text{grad}(x) \ln |x - y|]\varphi(y)\, ds(y) = (\text{grad } v)(x), \quad x \in S_0 \setminus \partial S,$$

$$F_0(x) = -\int_{\partial S} [\text{grad}(x) \ln |x - y|]\varphi(y)\, ds(y) = (\text{grad } v)_0(x), \quad x \in \partial S,$$

the latter understood as principal value.

1.49. Remark. Theorem 1.48 implies that if $\varphi \in C^{0,\alpha}(\partial S)$, then, as $S^{\pm} \ni x' \to x \in \partial S$, $\operatorname{grad} v(\varphi)$ tends to finite limits given by

$$(\operatorname{grad} v)^{\pm} = \pm \pi \nu(x) \varphi(x) - \int_{\partial S} [\operatorname{grad}(x) \ln |x - y|] \varphi(y) \, ds(y), \quad x \in \partial S, \tag{1.44}$$

where the last term is understood as principal value.

1.50. Remark. Theorems 1.48 and 1.17 also imply that if $\varphi \in C^{0,\alpha}(\partial S)$, $\alpha \in (0,1]$, then the restrictions of $v(\varphi)$ to $\bar{S}^+$ and $\bar{S}^-$ belong respectively to $C^{1,\beta}(\bar{S}^+)$ and $C^{1,\beta}(\bar{S}^-)$, with $\beta = \alpha$ for $\alpha \in (0,1)$ and any $\beta \in (0,1)$ for $\alpha = 1$. We denote these extensions by v^+ and v^-. Hence,

$$(\operatorname{grad} v^+)(x) = (\operatorname{grad} v)^+(x), \quad x \in \bar{S}^+,$$
$$(\operatorname{grad} v^-)(x) = (\operatorname{grad} v)^-(x), \quad x \in \bar{S}^-.$$

1.51. Theorem. *If $\varphi \in C^{1,\alpha}(\partial S)$, $\alpha \in (0,1]$, then the restrictions of $w(\varphi)$ to S^+ and S^- have $C^{1,\beta}$-extensions w^+ and w^- to $\bar{S}^+$ and $\bar{S}^-$, respectively, with $\beta = \alpha$ for $\alpha \in (0,1)$ and any $\beta \in (0,1)$ for $\alpha = 1$. These extensions are given by (1.37) and satisfy the equality $\partial w^+/\partial \nu = \partial w^-/\partial \nu$ on ∂S.*

Proof. Let $x \neq y$. Since

$$\Delta(y) \ln |x - y| = 0,$$
$$[\operatorname{grad}(x) + \operatorname{grad}(y)] \ln |x - y| = 0,$$

we can write

$$\frac{\partial}{\partial x_\gamma} \left[\frac{\partial}{\partial \nu(y)} \ln |x - y| \right]$$
$$= \nu_\beta(y) \frac{\partial}{\partial y_\beta} \left[\frac{\partial}{\partial x_\gamma} \ln |x - y| \right] + \nu_\gamma(y) \Delta(y) \ln |x - y|$$

$$= \nu_\beta(y)\frac{\partial}{\partial y_\beta}\left(-\frac{\partial}{\partial y_\gamma}\ln|x-y|\right) + \nu_\gamma(y)\frac{\partial}{\partial y_\beta}\left(\frac{\partial}{\partial y_\beta}\ln|x-y|\right)$$

$$= \left[\nu_\beta(y)\frac{\partial}{\partial y_\gamma} - \nu_\gamma(y)\frac{\partial}{\partial y_\beta}\right]\left(\frac{\partial}{\partial x_\beta}\ln|x-y|\right)$$

$$= \varepsilon_{\beta\gamma}\frac{\partial}{\partial s(y)}\left(\frac{\partial}{\partial x_\beta}\ln|x-y|\right).$$

Consequently, using integration by parts, we find that for $x \in S_0 \setminus \partial S$

$$\frac{\partial}{\partial x_\gamma}w(x) = -\int_{\partial S}\frac{\partial}{\partial x_\gamma}\left[\frac{\partial}{\partial \nu(y)}\ln|x-y|\right]\varphi(y)\,ds(y)$$

$$= \varepsilon_{\beta\gamma}\frac{\partial}{\partial x_\beta}\int_{\partial S}(\ln|x-y|)\varphi'(y)\,ds(y)$$

$$= \varepsilon_{\gamma\beta}\frac{\partial}{\partial x_\beta}(v(\varphi'))(x). \tag{1.45}$$

From this, Theorem 1.48 and the fact that $\varphi' \in C^{0,\alpha}(\partial S)$ we deduce that $\operatorname{grad} w(\varphi)$ has $C^{0,\beta}$-extensions $(\operatorname{grad} w)^+$ and $(\operatorname{grad} w)^-$ to $\bar{S}^+$ and $\bar{S}^-$. By Theorem 1.46, the extensions w^+ and w^-, given by (1.37), of $w(\varphi)$ are Hölder continuous on $\bar{S}^+$ and $\bar{S}^-$, respectively. Since $\operatorname{grad} w^+(x) = (\operatorname{grad} w)^+(x)$, $x \in S^+$, and $\operatorname{grad} w^-(x) = (\operatorname{grad} w)^-(x)$, $x \in S^-$, the first part of the assertion follows from Theorem 1.17.

To complete the proof we remark that, in view of (1.45) and (1.44), for $x \in \partial S$ Theorem 1.17 yields

$$\frac{\partial}{\partial \nu}w^\pm(x) = \langle\operatorname{grad} w^\pm(x), \nu(x)\rangle = \langle(\operatorname{grad} w)^\pm(x), \nu(x)\rangle$$

$$= \varepsilon_{\gamma\beta}\left[\frac{\partial}{\partial x_\beta}v(\varphi')\right]^\pm(x)\nu_\gamma(x)$$

$$= \varepsilon_{\beta\gamma}\nu_\gamma(x)\int_{\partial S}\left(\frac{\partial}{\partial x_\beta}\ln|x-y|\right)\varphi'(y)\,ds(y),$$

where the integral is understood as principal value.

1.52. Theorem. *The function w_0 defined by (1.38) as the direct values on ∂S of the double layer potential with density $\varphi \in C^{0,\alpha}(\partial S)$, $\alpha \in (0,1]$, belongs to $C^{1,\beta}(\partial S)$, with $\beta = \alpha$ for $\alpha \in (0,1)$ and any $\beta \in (0,1)$ for $\alpha = 1$.*

Proof. As noted in the proof of Theorem 1.46, the kernel $k(x,y) = -\partial \ln|x-y|/\partial \nu(y)$ is 0-singular on ∂S, consequently, $w_0(x)$ is an improper integral for all $x \in \partial S$. Clearly,

$$k_0(x,y) = \frac{\partial}{\partial s(x)} k(x,y)$$
$$= \frac{\langle \nu(y), \tau(x) \rangle}{|x-y|^2} - 2 \frac{\langle \nu(y), x-y \rangle \langle \tau(x), x-y \rangle}{|x-y|^4} \qquad (1.46)$$

is 1-singular on ∂S. Verifying the conditions of Lemma 1.25, we deduce that $k_0(x,y)$ is a proper 1-singular kernel on ∂S.

Next, by writing $\langle \cdot, \cdot \rangle$ in terms of the cosine of the angle between the vectors, we find that

$$\langle \nu(y), \tau(x) \rangle + \langle \nu(x), \tau(y) \rangle = 0, \quad x, y \in \partial S. \qquad (1.47)$$

Using the same technique, (1.46) and (1.47), for $x, y \in \partial S$, $x \neq y$, we now obtain

$$\left[\frac{\partial}{\partial s(x)} + \frac{\partial}{\partial s(y)} \right] k(x,y)$$
$$= 2|x-y|^{-4} \{ \langle \nu(x), x-y \rangle \langle \tau(y), x-y \rangle$$
$$- \langle \nu(y), x-y \rangle \langle \tau(x), x-y \rangle + \langle \nu(y), \tau(x) \rangle \} = 0. \qquad (1.48)$$

From this and (1.46) we conclude that $k_0(x,y)$ satisfies (1.25).

The assertion now follows from Theorem 1.34 with $\beta = \gamma = 0$ and $g(x) = -\pi$, $x \in \partial S$ (according to (1.42)), and Remarks 1.36 and 1.35.

1.6. Other potential-type functions

In this section we consider the Hölder continuity and continuous differentiability of some other useful integrals with γ-singular kernels.

1.53. Theorem. *Let $k(x, y)$ be continuous in $S_0 \times \partial S$ and such that $\mathrm{grad}(x)k(x, y)$ is a proper γ-singular kernel in S_0, $\gamma \in [0, 1)$, and let*

$$v^a(x) = \int_{\partial S} k(x, y)\varphi(y)\,ds(y), \quad x \in S_0.$$

If $\varphi \in C(\partial S)$, then $v^a \in C^{1,\beta}(S_0)$, with $\beta = 1 - \gamma$ for $\gamma \in (0, 1)$ and any $\beta \in (0, 1)$ for $\gamma = 0$.

Proof. Clearly, $v^a \in C(S_0) \cap C^1(S_0^+) \cap C^1(S_0^-)$. The statement follows from the fact that for $x \in S_0 \setminus \partial S$

$$\mathrm{grad}\, v^a(x) = \int_{\partial S} \mathrm{grad}(x)k(x, y)\varphi(y)\,ds(y),$$

which, by Theorem 1.30, belongs to $C^{0,\beta}(S_0)$.

1.54. Theorem. *Let $k(x, y)$ be continuous on $\partial S \times \partial S$ and such that $\partial k(x, y)/\partial s(x)$ is a proper γ-singular kernel on ∂S, $\gamma \in [0, 1)$. If $\varphi \in C(\partial S)$, then the function*

$$v_0^a(x) = \int_{\partial S} k(x, y)\varphi(y)\,ds(y), \quad x \in \partial S,$$

belongs to $C^{1,\beta}(\partial S)$, with $\beta = 1 - \gamma$ for $\gamma \in (0, 1)$ and any $\beta \in (0, 1)$ for $\gamma = 0$.

Proof. Consider the function
$$v_{0\delta}^a(x) = \int_{\partial S \setminus \Sigma_{x,\delta}} k(x,y)\varphi(y)\,ds(y), \quad \delta > 0.$$

It is obvious that $v_{0\delta}^a(x) \to v_0^a(x)$ as $\delta \to 0$, for all $x \in \partial S$. On the other hand,
$$\frac{\partial}{\partial s} v_{0\delta}^a(x) = \int_{\partial S \setminus \Sigma_{x,\delta}} \frac{\partial}{\partial s(x)} k(x,y)\varphi(y)\,ds(y),$$

which converges uniformly to $\int_{\partial S} [\partial k(x,y)/\partial s(x)]\varphi(y)\,ds(y)$ as $\delta \to 0$ (see the proof of Theorem 1.29). By a well-known theorem of analysis, v_0^a is differentiable at all $x \in \partial S$ and
$$\frac{\partial}{\partial s} v_0^a(x) = \int_{\partial S} \frac{\partial}{\partial s(x)} k(x,y)\varphi(y)\,ds(y).$$

We complete the proof by applying Theorem 1.30 to the above integral to deduce that $\partial v_0^a/\partial s \in C^{0,\beta}(\partial S)$.

1.55. Theorem. *If $\varphi \in C(\partial S)$, then the functions*
$$v_{\gamma\delta}^b(x) = \int_{\partial S} \frac{(x_\gamma - y_\gamma)(x_\delta - y_\delta)}{|x-y|^2} \varphi(y)\,ds(y), \quad x \in S_0, \tag{1.49}$$

$$v_\gamma^c(x) = \int_{\partial S} \left[\frac{\partial}{\partial s(y)}((x_\gamma - y_\gamma)\ln|x-y|)\right] \varphi(y)\,ds(y), \quad x \in S_0, \tag{1.50}$$

$$v_\gamma^d(x) = \int_{\partial S} \left[\frac{\partial}{\partial \nu(y)}((x_\gamma - y_\gamma)\ln|x-y|)\right] \varphi(y)\,ds(y), \quad x \in S_0, \tag{1.51}$$

belong to $C^{0,\alpha}(\mathbf{R}^2)$ for any $\alpha \in (0,1)$.

Proof. By direct verification or by means of Lemma 1.25, we easily convince ourselves that $(x_\gamma - y_\gamma)(x_\delta - y_\delta)|x-y|^{-2}$ is a proper 0-singular kernel in S_0. Similarly,

$$\frac{\partial}{\partial s(y)}[(x_\gamma - y_\gamma)\ln|x-y|] = -\tau_\gamma \ln|x-y| - \frac{(x_\gamma - y_\gamma)\langle \tau(y), x-y\rangle}{|x-y|^2}$$

and

$$\frac{\partial}{\partial \nu(y)}[(x_\gamma - y_\gamma)\ln|x-y|] = -\nu_\gamma(y)\ln|x-y| - \frac{(x_\gamma - y_\gamma)\langle \nu(y), x-y\rangle}{|x-y|^2}$$

are proper δ-singular kernels in S_0 for any $\delta \in (0,1)$. The result now follows from Theorem 1.30.

1.56. Theorem. *If $\varphi \in C^{0,\alpha}(\partial S)$, $\alpha \in (0,1]$, then the function*

$$v^e_{\gamma\delta}(x) = \int_{\partial S}\left[\frac{\partial}{\partial s(y)}\frac{(x_\gamma - y_\gamma)(x_\delta - y_\delta)}{|x-y|^2}\right]\varphi(y)\,ds(y), \quad x \in S_0, \quad (1.52)$$

belongs to $C^{0,\beta}(\mathbf{R}^2)$, with $\beta = \alpha$ for $\alpha \in (0,1)$ and any $\beta \in (0,1)$ for $\alpha = 1$.

Proof. Direct verification of the properties in Definition 1.23 shows that the kernel $k(x,y)$ of $v^e_{\gamma\delta}$ is a proper 1-singular kernel in S_0. Also, for $x, y \in \partial S$, $x \neq y$,

$$\frac{\partial}{\partial s(y)}\frac{(x_\gamma - y_\gamma)(x_\delta - y_\delta)}{|x-y|^2}$$

$$= c_{\gamma\delta\rho\sigma}\frac{(x_\rho - y_\rho)(x_\sigma - y_\sigma)}{|x-y|^2}\frac{\partial}{\partial \nu(y)}\ln|x-y|, \quad (1.53)$$

where

$$c_{\gamma\gamma\gamma\delta} = c_{\gamma\gamma\delta\gamma} = -c_{\gamma\delta\gamma\gamma} = -c_{\delta\gamma\gamma\gamma} = \varepsilon_{\delta\gamma},$$
$$c_{\gamma\delta\gamma\delta} = c_{\gamma\delta\delta\gamma} = c_{\gamma\gamma\delta\delta} = 0 \quad (\gamma, \delta \text{ not summed}),$$
$$(1.54)$$

which means that $k(x,y)$ is 0-singular on ∂S. Consequently, $v^e_{\gamma\delta}$ is an improper integral for $x \in \partial S$.

Since for $x, y \in \partial S$, $x \neq y$,

$$\lim_{y \to x} \frac{(x_\gamma - y_\gamma)(x_\delta - y_\delta)}{|x-y|^2} = \tau_\gamma(x)\tau_\delta(x),$$

we find that f and f_0 defined by (1.29) are identically zero. Hence, f^+ and f^- defined by (1.31) with $l(x) = 0$, $x \in \partial S$, belong to $C^{0,\alpha}(\partial S)$. The result now follows from Theorem 1.40.

1.57. Theorem. *If $\varphi \in C^{0,\alpha}(\partial S)$, $\alpha \in (0,1]$, then*

$$v_0^f(x) = \int_{\partial S} \left[\frac{\partial}{\partial s(y)} \ln|x-y| \right] \varphi(y)\, ds(y), \quad x \in \partial S, \qquad (1.55)$$

exists as principal value uniformly for all $x \in \partial S$. Furthermore, $v_0^f \in C^{0,\beta}(\partial S)$, with $\beta = \alpha$ for $\alpha \in (0,1)$ and any $\beta \in (0,1)$ for $\alpha = 1$.

Proof. For $x, y \in \partial S$, $x \neq y$, we have

$$\left| \frac{\partial}{\partial s(y)} \ln|x-y| \right| = \frac{|\langle \tau(y), x-y \rangle|}{|x-y|^2} \leq c_1 |x-y|^{-1},$$

$$|x-y| \left| \frac{\partial}{\partial x_\gamma} \left[\frac{\partial}{\partial s(y)} \ln|x-y| \right] \right|$$

$$= \left| \frac{\tau_\gamma(y)}{|x-y|} - 2\frac{\langle \tau(y), x-y \rangle (x_\gamma - y_\gamma)}{|x-y|^3} \right| \leq c_2 |x-y|^{-1},$$

where c_1 and c_2 are positive constants. Therefore, by Lemma 1.25, $\partial \ln|x-y|/\partial s(y)$ is a proper 1-singular kernel on ∂S. This kernel is also uniformly integrable since if a and b are the end-points of $\Sigma_{x,\delta}$, then

$$\int_{\partial S \setminus \Sigma_{x,\delta}} \frac{\partial}{\partial s(y)} \ln|x-y|\, ds(y) = \ln \frac{|x-a|}{|x-b|} = 0 \qquad (1.56)$$

for all $0 < \delta \le r$ and all $x \in \partial S$. We can now write

$$\int_{\partial S \setminus \Sigma_{x,\delta}} \left[\frac{\partial}{\partial s(y)} \ln |x-y|\right] \varphi(y)\, ds(y)$$

$$= \int_{\partial S \setminus \Sigma_{x,\delta}} \left[\frac{\partial}{\partial s(y)} \ln |x-y|\right] [\varphi(y) - \varphi(x)]\, ds(y),$$

and the first part of the assertion follows from Definition 1.37 and the uniform convergence, as $\delta \to 0$, of the right-hand side, whose integrand is $O(|x-y|^{\alpha-1})$. Consequently,

$$v_0^f(x) = \int_{\partial S} \left[\frac{\partial}{\partial s(y)} \ln |x-y|\right] [\varphi(y) - \varphi(x)]\, ds(y), \quad x \in \partial S, \qquad (1.57)$$

in the sense of principal value.

To compete the proof we apply Theorem 1.32 with $\xi \equiv x$ and make use of the last part of Remark 1.42.

1.58. Theorem. *If $\varphi \in C^{0,\alpha}(\partial S)$, $\alpha \in (0,1]$, then the function*

$$v^f(x) = \int_{\partial S} \left[\frac{\partial}{\partial s(y)} \ln |x-y|\right] \varphi(y)\, ds(y), \quad x \in S_0 \setminus \partial S, \qquad (1.58)$$

is $C^{0,\beta}$-extendable to $\mathbf{R}^2$, with $\beta = \alpha$ for $\alpha \in (0,1)$ and any $\beta \in (0,1)$ for $\alpha = 1$.

Proof. In the proof of Theorem 1.57 it was shown that $k(x,y) = \partial \ln|x-y|/\partial s(y)$ is an integrable, proper 1-singular kernel on ∂S. The same reasoning indicates that $k(x,y)$ is also a proper 1-singular kernel in S_0. In view of (1.56), the formulae (1.29) yield

$$\begin{aligned} f(x) &= 0, \quad x \in S_0 \setminus \partial S, \\ f_0(x) &= 0, \quad x \in \partial S, \end{aligned} \qquad (1.59)$$

the latter understood as principal value. From (1.59) and (1.31) with $l(x) = 0$, $x \in \partial S$, it follows that $f^+ \in C^{0,\alpha}(\bar{S}_0^+)$ and $f^- \in C^{0,\alpha}(\bar{S}_0^-)$ (both these functions are identically zero). The application of Theorem 1.44 now completes the proof.

1.59. Remark. Since $l = 0$, (1.58) also represents the extension of v^f to $\mathbf{R}^2$, that is, it holds for $x \in \mathbf{R}^2$, but for $x \in \partial S$ the integral on the right-hand side (denoted by v_0^f in (1.55)) must be understood as principal value.

Alternatively, since

$$\int_{\partial S} \frac{\partial}{\partial s(y)} \ln|x-y|\, ds(y) = 0, \quad x \in \mathbf{R}^2 \setminus \partial S,$$

we see that the extension of v^f to $\mathbf{R}^2$ is also given by the right-hand side of (1.57) with $x \in \mathbf{R}^2$.

1.60. Theorem. *If $\varphi \in C^{1,\alpha}(\partial S)$, $\alpha \in (0,1]$, then the function v_0^f defined by (1.55) belongs to $C^{1,\beta}(\partial S)$, with $\beta = \alpha$ for $\alpha \in (0,1)$ and any $\beta \in (0,1)$ for $\alpha = 1$.*

Proof. By Theorem 1.57, v_0^f is Hölder continuous on ∂S.

Let $x = \psi(s) \in \partial S$ be arbitrary but fixed, and let $a = \psi(s-\delta)$ and $b = \psi(s+\delta)$ be the end-points of the arc $\Gamma_{x,\delta}$ defined by (1.28). Integrating by parts, we find that

$$\int_{\partial S \setminus \Gamma_{x,\delta}} (\ln|x-y|)\varphi'(y)\, ds(y) = \varphi(a)\ln|x-a| - \varphi(b)\ln|x-b|$$

$$- \int_{\partial S \setminus \Gamma_{x,\delta}} \left[\frac{\partial}{\partial s(y)} \ln|x-y|\right] \varphi(y)\, ds(y). \quad (1.60)$$

The first term on the right-hand side can be written in the form

$$\varphi(x)(\ln |x - a| - \ln |x - b|)$$
$$+ [\varphi(a) - \varphi(x)] \ln |x - a| - [\varphi(b) - \varphi(x)] \ln |x - b|.$$

Since

$$\ln |x - a| - \ln |x - b| = \ln \left(\frac{|x - a|}{\delta} \cdot \frac{\delta}{|x - b|} \right)$$

and φ is differentiable on ∂S, by Theorem 1.14 this expression tends to zero as $\delta \to 0$.

In the proof of Theorem 1.58 it was shown that $\partial \ln |x - y|/\partial s(y)$ is an integrable, proper 1-singular kernel on ∂S. Setting

$$F(x) = \int_{\partial S} (\ln |x - y|) \varphi'(y) \, ds(y),$$

$$F_\delta(x) = \int_{\partial S \setminus \Gamma_{x,\delta}} (\ln |x - y|) \varphi'(y) \, ds(y),$$

and letting $\delta \to 0$ in (1.60), we see that, by Theorem 1.57 and Remark 1.38,

$$F(x) = \lim_{\delta \to 0} F_\delta(x) = -v_0^f(x). \tag{1.61}$$

On the other hand, by Leibniz's rule for differentiating an integral whose limits depend on the parameter,

$$F'_\delta(x) = \int_{\partial S \setminus \Gamma_{x,\delta}} \left[\frac{\partial}{\partial s(x)} \ln |x - y| \right] \varphi'(y) \, ds(y)$$
$$+ \varphi'(a) \ln |x - a| - \varphi'(b) \ln |x - b|.$$

Since $\varphi' \in C^{0,\alpha}(\partial S)$, we deduce as above that the sum of the last two terms tends to zero uniformly as $\delta \to 0$. Hence,

$$\lim_{\delta \to 0} F'_\delta(x) = \int_{\partial S} \left[\frac{\partial}{\partial s(x)} \ln |x - y| \right] \varphi'(y) \, ds(y), \tag{1.62}$$

where the integral is understood as principal value and, by Theorem 1.57, the convergence is uniform with respect to x. A well-known result of analysis now implies that $F(x)$ is differentiable and that $F'(x)$ is equal to the right-hand side of (1.62). Taking (1.61) into account, we conclude that $\partial v_0^f(x)/\partial s$ exists and

$$\frac{\partial}{\partial s} v_0^f(x) = -\int_{\partial S} \left[\frac{\partial}{\partial s(x)} \ln|x-y| \right] \varphi'(y)\, ds(y), \quad x \in \partial S.$$

By Theorem 1.57, $\partial v_0^f/\partial s \in C^{0,\alpha}(\partial S)$, as required.

1.61. Theorem. *If $\varphi \in C^{0,\alpha}(\partial S)$, $\alpha \in (0,1]$, then the functions*

$$v_{\gamma 0}^c(x) = \int_{\partial S} \left\{ \frac{\partial}{\partial s(y)} [(x_\gamma - y_\gamma) \ln|x-y|] \right\} \varphi(y)\, ds(y), \quad x \in \partial S, \quad (1.63)$$

$$v_{\gamma 0}^d(x) = \int_{\partial S} \left\{ \frac{\partial}{\partial \nu(y)} [(x_\gamma - y_\gamma) \ln|x-y|] \right\} \varphi(y)\, ds(y), \quad x \in \partial S, \quad (1.64)$$

belong to $C^{1,\beta}(\partial S)$, with $\beta = \alpha$ for $\alpha \in (0,1)$ and any $\beta \in (0,1)$ for $\alpha = 1$.

Proof. The kernel

$$k(x, y) = \frac{\partial}{\partial s(y)} [(x_\gamma - y_\gamma) \ln|x-y|]$$

$$= -\tau_\gamma(y) \ln|x-y| - \frac{(x_\gamma - y_\gamma)\langle \tau(y), x-y \rangle}{|x-y|^2}$$

is δ-singular on ∂S, where $\delta \in (0,1)$ is arbitrary, so $v_{\gamma 0}^c(x)$ is an improper integral for all $x \in \partial S$. Also,

$$k_0(x, y) = \frac{\partial}{\partial s(x)} k(x, y)$$

$$= -\tau_\gamma(y) \frac{\partial}{\partial s(x)} \ln|x-y| - \frac{\partial}{\partial s(x)} \frac{(x_\gamma - y_\gamma)\langle \tau(y), x-y \rangle}{|x-y|^2} \quad (1.65)$$

is 1-singular on ∂S. Using Lemma 1.25, we find that $k_0(x, y)$ is a proper 1-singular kernel on ∂S.

Since
$$\hat{k}(x, y) = \left[\frac{\partial}{\partial s(x)} + \frac{\partial}{\partial s(y)}\right] \ln|x - y| = \frac{\langle \tau(x) - \tau(y), x - y\rangle}{|x - y|^2}$$

is 0-singular on ∂S, the first term on the right-hand side in (1.65) can be written in the form
$$-\hat{k}(x, y)\tau_\gamma(y) + \frac{\partial}{\partial s(y)}[\tau_\gamma(y) \ln|x - y|] - \kappa(y)\nu_\gamma(y)\ln|x - y|.$$

Similarly, since
$$\tilde{k}_{\gamma\delta}(x, y) = \left[\frac{\partial}{\partial s(x)} + \frac{\partial}{\partial s(y)}\right]\frac{(x_\gamma - y_\gamma)(x_\delta - y_\delta)}{|x - y|^2}$$
$$= \frac{[\tau_\gamma(x) - \tau_\gamma(y)](x_\delta - y_\delta)}{|x - y|^2} + \frac{[\tau_\delta(x) - \tau_\delta(y)](x_\gamma - y_\gamma)}{|x - y|^2}$$
$$- 2\frac{(x_\gamma - y_\gamma)(x_\delta - y_\delta)\langle\tau(x) - \tau(y), x - y\rangle}{|x - y|^4}$$

is 0-singular on ∂S, the second term on the right-hand side in (1.65) becomes
$$-\tilde{k}_{\gamma\delta}(x, y)\tau_\delta(y) + \frac{\partial}{\partial s(y)}\frac{(x_\gamma - y_\gamma)\langle\tau(y), x - y\rangle}{|x - y|^2}$$
$$- \frac{\kappa(y)(x_\gamma - y_\gamma)\langle\nu(y), x - y\rangle}{|x - y|^2}.$$

Denoting by a and b the end-points of $\Sigma_{x,\delta}$, we find that
$$\int_{\partial S\setminus\Sigma_{x,\delta}} \frac{\partial}{\partial s(y)}[\tau_\gamma(y)\ln|x - y|]\,ds(y)$$
$$= \tau_\gamma(a)\ln|x - a| - \tau_\gamma(b)\ln|x - b|$$

$$= \tau_\gamma(a) \ln \frac{|x-a|}{|x-b|} + [\tau_\gamma(a) - \tau_\gamma(b)] \ln |x-b|$$

$$= [\tau_\gamma(a) - \tau_\gamma(b)] \ln |x-b| \to 0$$

uniformly as $\delta \to 0$, and that

$$\int_{\partial S \setminus \Sigma_{x,\delta}} \frac{\partial}{\partial s(y)} \frac{(x_\gamma - y_\gamma)\langle \tau(y), x-y \rangle}{|x-y|^2} ds(y)$$

$$= \frac{(x_\gamma - a_\gamma)\langle \tau(a), x-a \rangle}{|x-a|^2} - \frac{(x_\gamma - b_\gamma)\langle \tau(b), x-b \rangle}{|x-b|^2} \to 0$$

uniformly as $\delta \to 0$. Consequently, the kernel $k_0(x, y)$ satisfies the estimate (1.25).

The result now follows from Theorem 1.34 with any $\beta \in (0, \alpha)$, $\gamma = 0$, and $g(x) = 0$, $x \in \partial S$, and Remarks 1.36 and 1.35.

The function $v_{\gamma 0}^d$ is treated similarly.

1.62. Theorem. *If $\varphi \in C^{0,\alpha}(\partial S)$, $\alpha \in (0, 1]$, then the function defined by*

$$v_{\gamma \delta 0}^e(x) = \int_{\partial S} \left[\frac{\partial}{\partial s(y)} \frac{(x_\gamma - y_\gamma)(x_\delta - y_\delta)}{|x-y|^2} \right] \varphi(y) \, ds(y), \quad x \in \partial S, \quad (1.66)$$

belongs to $C^{1,\beta}(\partial S)$, with $\beta = \alpha$ for $\alpha \in (0, 1)$ and any $\beta \in (0, 1)$ for $\alpha = 1$.

Proof. From the formula (1.53) and the estimates in Lemma 1.1 we see that the kernel

$$k(x, y) = \frac{\partial}{\partial s(y)} \frac{(x_\gamma - y_\gamma)(x_\delta - y_\delta)}{|x-y|^2}$$

is 0-singular on ∂S, hence $v_{\gamma \delta 0}^e(x)$ is an improper integral for all $x \in \partial S$.

A simple calculation shows that

$$
\begin{aligned}
k_0(x, y) &= \frac{\partial}{\partial s(x)} k(x, y) \\
&= c_{\gamma\delta\lambda\mu} c_{\lambda\mu\rho\sigma} \frac{(x_\rho - y_\rho)(x_\sigma - y_\sigma)}{|x - y|^2} \left[\frac{\partial}{\partial \nu(x)} \ln |x - y| \right] \left[\frac{\partial}{\partial \nu(y)} \ln |x - y| \right] \\
&\quad + c_{\gamma\delta\rho\sigma} \frac{(x_\rho - y_\rho)(x_\sigma - y_\sigma)}{|x - y|^2} \frac{\partial}{\partial s(x)} \left[\frac{\partial}{\partial \nu(y)} \ln |x - y| \right],
\end{aligned} \quad (1.67)
$$

where the $c_{\gamma\delta\rho\sigma}$ are given by (1.54), is a 1-singular kernel on ∂S. Moreover, using Lemma 1.25, we easily convince ourselves that $k_0(x, y)$ is a proper 1-singular kernel on ∂S.

The first term on the right-hand side of (1.67) is 0-singular on ∂S. By (1.48), the second term can be written in the form

$$
c_{\gamma\delta\rho\sigma} \left\{ -\frac{\partial}{\partial s(y)} \left[\frac{(x_\rho - y_\rho)(x_\sigma - y_\sigma)}{|x - y|^2} \frac{\partial}{\partial \nu(y)} \ln |x - y| \right] \right. \\
\left. + \left[\frac{\partial}{\partial \nu(y)} \ln |x - y| \right] \frac{\partial}{\partial s(y)} \frac{(x_\rho - y_\rho)(x_\sigma - y_\sigma)}{|x - y|^2} \right\},
$$

from which, in view of what was said above about $k(x, y)$, we immediately deduce by direct verification that $k_0(x, y)$ satisfies the estimate (1.25).

The assertion now follows from Theorem 1.34 with $\beta = \gamma = 0$ and $g(x) = 0$, $x \in \partial S$, and Remarks 1.36 and 1.35.

1.7. Complex singular kernels

In the analysis of two-dimensional problems it is often convenient to express certain properties of functions in terms of complex variables. Extending an earlier convention, for a function f given on ∂S we write $f(z) \equiv f(x)$, where $z = x_1 + ix_2$.

Suppose now that $C(\partial S)$ and $C^1(\partial S)$ are complex vector spaces, and construct the complex spaces $C^{0,\alpha}(\partial S)$ and $C^{1,\alpha}(\partial S)$ by defining Hölder continuity in terms of the inequality

$$|f(z) - f(\zeta)| \leq c|z - \zeta|^\alpha \quad \text{for all} \quad z, \zeta \in \partial S,$$

and the derivative as

$$f'(z) = \frac{d}{dz} f(z) = \lim_{\zeta \to z} \frac{f(\zeta) - f(z)}{\zeta - z}, \quad z, \zeta \in \partial S,$$

if this limit exists.

Since $|z - \zeta| = |x - y|$, where $\zeta = y_1 + iy_2$, it is obvious that Hölder continuity with respect to z and Hölder continuity with respect to x (or s, according to the discussion in §1.2) are equivalent. The same can also be said about Hölder continuous differentiability on ∂S. We can see this from the equality

$$f'(z) = \vartheta(z) f'(s),$$

where

$$\vartheta(z) = \frac{dz}{ds} = \tau_1(z) + i\tau_2(z),$$

which implies that $f' \in C^{0,\alpha}(\partial S)$ in terms of z if and only if $f' \in C^{0,\alpha}(\partial S)$ in terms of s, in view of Lemma 1.20 and the fact that both $\vartheta(z)$ and $[\vartheta(z)]^{-1} = \bar\vartheta(z) = \tau_1(z) - i\tau_2(z)$ belong to $C^1(\partial S)$. This shows that our somewhat loose use of the same symbol for a function on ∂S whether it is expressed in terms of z or x is justified in relation to Hölder spaces.

In the light of these arguments, and because for a kernel $k(x, y)$ and a density φ on ∂S

$$\int_{\partial S} k(x, y) \varphi(y) \, ds(y) = \int_{\partial S} k(z, \zeta) \varphi(\zeta) \bar\vartheta(\zeta) \, d\zeta,$$

we conclude that the definition of γ-singular and proper γ-singular kernels on ∂S and all the associated results established in §1.4 on the behaviour on ∂S of integrals with such kernels can be understood in terms of either real or complex variables.

1.63. Theorem. *If $\varphi \in C^{0,\alpha}(\partial S)$, $\alpha \in (0,1]$, then*

$$\Psi(z) = \int_{\partial S} \frac{\varphi(\zeta)}{\zeta - z} d\zeta, \quad z \in \partial S, \tag{1.69}$$

exists in the sense of principal value, uniformly for all $z \in \partial S$, and belongs to $C^{0,\beta}(\partial S)$, with $\beta = \alpha$ for $\alpha \in (0,1)$ and any $\beta \in (0,1)$ for $\alpha = 1$.

Proof. Let $z = x_1 + ix_2$ and $\zeta = y_1 + iy_2$. Differentiating with respect to $s(y)$ the equality

$$\log(\zeta - z) = \ln|\zeta - z| + i\theta = \ln|x - y| + i\theta,$$

where $\theta = \arg(\zeta - z)$, and using the Cauchy-Riemann relation

$$\frac{\partial}{\partial s(y)} \theta(x, y) = \frac{\partial}{\partial \nu(y)} \ln|x - y|,$$

we obtain

$$\frac{d\zeta}{\zeta - z} = \frac{\partial}{\partial s(y)} \ln|x - y| \, ds(y) + i \frac{\partial}{\partial \nu(y)} \ln|x - y| \, ds(y). \tag{1.70}$$

Hence, we can write

$$\int_{\partial S \setminus \Sigma_{x,\delta}} \frac{\varphi(\zeta)}{\zeta - z} d\zeta = \int_{\partial S \setminus \Sigma_{x,\delta}} \left[\frac{\partial}{\partial s(y)} \ln|x - y| \right] \varphi(y) \, ds(y)$$

$$+ i \int_{\partial S \setminus \Sigma_{x,\delta}} \left[\frac{\partial}{\partial \nu(y)} \ln|x - y| \right] \varphi(y) \, ds(y),$$

and the result is obtained from Theorems 1.57 and 1.46 by letting $\delta \to 0$.

1.64. Remark. The function Ψ defined by (1.69) can be expressed in terms of an improper integral. Writing

$$\int_{\partial S\setminus\Sigma_{x,\delta}} \frac{\varphi(\zeta)}{\zeta - z}\, d\zeta = \int_{\partial S\setminus\Sigma_{x,\delta}} \frac{\varphi(\zeta) - \varphi(z)}{\zeta - z}\, d\zeta + \varphi(z) \int_{\partial S\setminus\Sigma_{x,\delta}} \frac{d\zeta}{\zeta - z},$$

replacing $(\zeta - z)^{-1} d\zeta$ by its expression in (1.70), letting $\delta \to 0$, and using the formulae (1.56) and (1.42), we find that, in the sense of principal value,

$$\int_{\partial S} \frac{\varphi(\zeta)}{\zeta - z}\, d\zeta = \pi i \varphi(z) + \int_{\partial S} \frac{\varphi(\zeta) - \varphi(z)}{\zeta - z}\, d\zeta, \quad z \in \partial S, \tag{1.71}$$

where the integrand of the last term is $O(|z - \zeta|^{\alpha - 1})$ if $\varphi \in C^{0,\alpha}(\partial S)$, $\alpha \in (0, 1]$.

1.65. Theorem. *If $\varphi \in C^{1,\alpha}(\partial S)$, $\alpha \in (0,1]$, then Ψ defined by (1.69) belongs to $C^{1,\beta}(\partial S)$, with $\beta = \alpha$ for $\alpha \in (0,1)$ and any $\beta \in (0,1)$ for $\alpha = 1$.*

Proof. By (1.70), (1.55) and (1.36),

$$\Psi(z) = v_0^f(x) - iw(x),$$

and the assertion follows from Theorems 1.60 and 1.52.

1.66. Theorem. *If $K^s : C^{0,\alpha}(\partial S) \to C^{0,\alpha}(\partial S)$, $\alpha \in (0,1)$, is the operator defined by*

$$K^s(\varphi)(z) = \int_{\partial S} \frac{\varphi(\zeta)}{\zeta - z}\, d\zeta, \quad z \in \partial S, \tag{1.72}$$

then $(K^s)^2 = -\pi^2 I$, where I is the identity operator.

Proof. From Theorem 1.63 it is clear that the operator composition $(K^s)^2$ is meaningful.

In [27] it is shown that a function $f(z, \zeta)$ which is Hölder continuous with respect to both its variables z and ζ satisfies the Poincaré-Bertrand formula

$$\int_{\partial S} \frac{1}{\zeta - z} \left[\int_{\partial S} \frac{f(\zeta, \eta)}{\eta - \zeta} d\eta \right] d\zeta$$
$$= -\pi^2 f(z, z) + \int_{\partial S} \int_{\partial S} \frac{f(\zeta, \eta)}{(\zeta - z)(\eta - \zeta)} d\zeta \, d\eta. \quad (1.73)$$

Using (1.73) and the fact that, by (1.71) with $\varphi = 1$,

$$\int_{\partial S} \frac{d\zeta}{\zeta - z} = \pi i, \quad z \in \partial S, \quad (1.74)$$

in the sense of principal value, we find that for any $\varphi \in C^{0,\alpha}(\partial S)$ and $z \in \partial S$

$$((K^s)^2 \varphi)(z) = \int_{\partial S} \frac{1}{\zeta - z} \left[\int_{\partial S} \frac{\varphi(\eta)}{\eta - \zeta} d\eta \right] d\zeta$$
$$= -\pi^2 \varphi(z) + \int_{\partial S} \int_{\partial S} \frac{\varphi(\eta)}{(\zeta - z)(\eta - \zeta)} d\zeta \, d\eta$$
$$= -\pi^2 \varphi(z) + \int_{\partial S} \left[\frac{1}{\eta - z} \left(\int_{\partial S} \frac{d\zeta}{\zeta - z} - \int_{\partial S} \frac{d\zeta}{\zeta - \eta} \right) \varphi(\eta) \right] d\eta$$
$$= -\pi^2 \varphi(z),$$

as required.

1.67. Theorem. *Let $f(z, \zeta)$ be a function defined on $\partial S \times \partial S$, which belongs to $C^{0,\alpha}(\partial S)$, $\alpha \in (0, 1]$, with respect to each of its variables, uniformly relative to the other one, and satisfies the inequality*

$$|f(z, \zeta) - f(z', \zeta)| < c|z - z'| |z - \zeta|^{\alpha - 1}, \quad c = \text{const} > 0,$$

for all $z, z', \zeta \in \partial S$ such that $0 < |z - z'| < \frac{1}{2}|z - \zeta|$. Then the function

$$\Lambda(z) = \int_{\partial S} \frac{f(z, \zeta)}{\zeta - z} d\zeta, \quad z \in \partial S,$$

where the integral is understood as principal value, belongs to $C^{0,\beta}(\partial S)$, with $\beta = \alpha$ for $\alpha \in (0, 1)$ and any $\beta \in (0, 1)$ for $\alpha = 1$.

Proof. Let $z = x_1 + ix_2$. Writing

$$\int_{\partial S \setminus \Sigma_{x,r}} \frac{f(z, \zeta)}{\zeta - z} d\zeta = \int_{\partial S \setminus \Sigma_{x,r}} \frac{f(z, \zeta) - f(z, z)}{\zeta - z} d\zeta + f(z, z) \int_{\partial S \setminus \Sigma_{x,r}} \frac{d\zeta}{\zeta - z},$$

from Theorem 1.63 and the fact that the integrand of the first term on the right-hand side is $O(|z - \zeta|^{\alpha-1})$ we conclude that $\Lambda(z)$ exists in the sense of principal value for all $z \in \partial S$.

To establish the Hölder continuity of Λ, for $z, z', \zeta \in \partial S$ we use the decomposition

$$2[\Lambda(z) - \Lambda(z')] = \int_{\partial S} \left[\frac{f(z, \zeta) - f(z, z)}{\zeta - z} - \frac{f(z, \zeta) - f(z, z')}{\zeta - z'} \right] d\zeta$$

$$+ \int_{\partial S} \left[\frac{f(z', \zeta) - f(z', z)}{\zeta - z} - \frac{f(z', \zeta) - f(z', z')}{\zeta - z'} \right] d\zeta$$

$$+ \int_{\partial S} \frac{f(z, \zeta) - f(z', \zeta)}{\zeta - z} d\zeta + \int_{\partial S} \frac{f(z, \zeta) - f(z', \zeta)}{\zeta - z'} d\zeta$$

$$+ f(z, z) \int_{\partial S} \frac{d\zeta}{\zeta - z} - f(z, z') \int_{\partial S} \frac{d\zeta}{\zeta - z'}$$

$$+ f(z', z) \int_{\partial S} \frac{d\zeta}{\zeta - z} - f(z', z') \int_{\partial S} \frac{d\zeta}{\zeta - z'}$$

$$= I_1 + I_2 + I_3 + I_4 + I_5 + I_6.$$

Let $z' = x_1' + ix_2'$ and $\zeta = y_1 + iy_2$, and let $\Sigma_{x,r}$, Σ_1 and Σ_2 be the sets defined by (1.12), (1.15) and (1.16), with x and x' satisfying (1.18). By Lemmas 1.8–1.11, (1.74) and Remark 1.38,

$$|I_{11}| = \left| \int_{\Sigma_1} \left[\frac{f(z,\zeta) - f(z,z)}{\zeta - z} - \frac{f(z,\zeta) - f(z,z')}{\zeta - z'} \right] d\zeta \right|$$

$$\leq c_1 \int_{\Sigma_1} (|x-y|^{\alpha-1} + |x'-y|^{\alpha-1}) \, ds(y) \leq c_2 |z-z'|^\alpha,$$

$$|I_{12}| = \left| \int_{\Sigma_2} \left(\frac{1}{\zeta - z} - \frac{1}{\zeta - z'} \right) [f(z,\zeta) - f(z,z')] \, d\zeta \right|$$

$$\leq c_3 |z-z'| \int_{\Sigma_2} |x-y|^{\alpha-2} ds(y) \leq c_4 |z-z'|^\alpha \quad \text{if } \alpha \in (0,1),$$

$$|I_{12}| \leq c_5 |z-z'|^\alpha |\ln |z-z'|| \quad \text{if } \alpha = 1,$$

$$|I_{13}| = \left| \int_{\partial S \setminus \Sigma_{x,r}} \left(\frac{1}{\zeta - z} - \frac{1}{\zeta - z'} \right) [f(z,\zeta) - f(z,z')] \, d\zeta \right|$$

$$\leq c_6 |z-z'| \int_{\partial S \setminus \Sigma_{x,r}} |x-y|^{\alpha-2} ds(y) \leq c_7 |z-z'|,$$

$$|I_{14}| = \left| [f(z,z') - f(z,z)] \int_{\partial S \setminus \Sigma_1} \frac{d\zeta}{\zeta - z} \right| \leq c_8 |z-z'|^\alpha,$$

consequently,

$$|I_1| = |I_{11} + I_{12} + I_{13} + I_{14}| \leq c_9 |z-z'|^\beta,$$

where the constants $c_1, \ldots, c_9 > 0$ may depend on α.

Similarly,

$$|I_2| \leq c_{10} |z-z'|^\beta, \quad c_{10} = \text{const} > 0.$$

Next, we find that

$$|I_{31}| = \left| \int_{\Sigma_1} \{[f(z,\zeta) - f(z,z)] - [f(z',\zeta) - f(z',z)]\} \frac{d\zeta}{\zeta - z} \right|$$

$$\leq c_{11} \int_{\Sigma_1} |x-y|^{\alpha-1} ds(y) \leq c_{12} |z-z'|^\alpha,$$

$$|I_{32}| = \left| \int_{\Sigma_2} \frac{f(z,\zeta) - f(z',\zeta)}{\zeta - z} d\zeta \right|$$

$$\leq c_{13}|z-z'| \int_{\Sigma_2} |x-y|^{\alpha-2} ds(y) \leq c_{14}|z-z'|^\alpha \quad \text{if} \quad \alpha \in (0,1),$$

$$|I_{32}| \leq c_{15}|z-z'| \, |\ln|z-z'|| \quad \text{if} \quad \alpha = 1,$$

$$|I_{33}| = \left| \int_{\partial S \setminus \Sigma_{x,r}} \frac{f(z,\zeta) - f(z',\zeta)}{\zeta - z} d\zeta \right|$$

$$\leq c_{16}|z-z'| \int_{\partial S \setminus \Sigma_{x,r}} |x-y|^{\alpha-2} ds(y) \leq c_{17}|z-z'|,$$

$$|I_{34}| = \left| [f(z,z) - f(z',z)] \left(\int_{\partial S} \frac{d\zeta}{\zeta - z} - \int_{\partial S \setminus \Sigma_1} \frac{d\zeta}{\zeta - z} \right) \right| \leq c_{18}|z-z'|^\alpha,$$

therefore,

$$|I_3| = |I_{31} + I_{32} + I_{33} + I_{34}| \leq c_{19}|z-z'|^\beta,$$

where the constants $c_{11}, \ldots, c_{19} > 0$ may depend on α. In exactly the same way, but using $\Sigma_{x',r}$ instead of $\Sigma_{x,r}$, we find that

$$|I_4| \leq c_{20}|z-z'|^\beta, \quad c_{20} = \text{const} > 0.$$

Finally,

$$|I_5| = |\pi i [f(z,z) - f(z,z')]| \leq c_{21}|z-z'|^\alpha,$$

$$|I_6| = |\pi i [f(z',z) - f(z',z')]| \leq c_{22}|z-z'|^\alpha,$$

where c_{21} and c_{22} are positive constants.

Combining the above inequalities, we now obtain

$$|\Lambda(z) - \Lambda(z')| \leq c_{23}|z - z'|^{\beta}, \quad c_{23} = \text{const} > 0,$$

as required.

1.8. Singular integral equations

We discuss briefly a few concepts of functional analysis, which will enable us to find the solutions of the boundary value problems to be stated later in §2.4. The presentation is made in terms of complex variables in order to simplify the technicalities involved. Any difference between the complex and real cases will be indicated explicitly.

1.68. Theorem. $C^{0,\alpha}(\partial S)$ *is a Banach space with norm*

$$\|\varphi\|_{\alpha} = \|\varphi\|_{\infty} + |\varphi|_{\alpha}, \qquad (1.75)$$

where

$$\|\varphi\|_{\infty} = \sup_{z \in \partial S} |\varphi(z)|,$$

$$|\varphi|_{\alpha} = \sup_{\substack{z,\zeta \in \partial S \\ z \neq \zeta}} \frac{|\varphi(z) - \varphi(\zeta)|}{|z - \zeta|^{\alpha}}.$$

Proof. As can easily be verified, (1.75) satisfies satisfies the norm axioms.

Let $\{\varphi_n\}_{n=1}^{\infty}$ be a Cauchy sequence in $C^{0,\alpha}(\partial S)$, that is, for any $\varepsilon > 0$ arbitrarily small there is a positive integer $n_0(\varepsilon)$ such that

$$\|\varphi_n - \varphi_m\|_{\alpha} < \varepsilon \quad \text{for all} \quad n, m > n_0(\varepsilon).$$

By (1.75),

$$\|\varphi_n - \varphi_m\|_{\infty} < \varepsilon \quad \text{for all} \quad n, m > n_0(\varepsilon),$$

which means that $\{\varphi_n\}_{n=1}^\infty$ is also a Cauchy sequence in $C(\partial S)$. Since $C(\partial S)$ is a complete space, there is a $\varphi \in C(\partial S)$ such that

$$\|\varphi_n - \varphi_m\|_\infty \to 0 \quad \text{as} \quad n \to \infty. \tag{1.76}$$

From (1.75) we also deduce that

$$|\varphi_n - \varphi_m|_\alpha < \varepsilon \quad \text{for all} \quad n, m > n_0(\varepsilon).$$

Letting $m \to \infty$ and using the uniform convergence of $\{\varphi_n\}_{n=1}^\infty$ on ∂S, we now obtain

$$|\varphi_n - \varphi|_\alpha < \varepsilon \quad \text{for all} \quad n > n_0(\varepsilon). \tag{1.77}$$

Hence, there is a $c = \text{const} > 0$ such that for all $z, \zeta \in \partial S$, $z \neq \zeta$,

$$\frac{|\varphi(z) - \varphi(\zeta)|}{|z - \zeta|^\alpha} \leq |\varphi|_\alpha \leq c,$$

in other words, $\varphi \in C^{0,\alpha}(\partial S)$. Also, from (1.76) and (1.77) it follows that

$$\|\varphi_n - \varphi\|_\alpha \to 0 \quad \text{as} \quad n \to \infty,$$

that is, $\{\varphi_n\}_{n=1}^\infty$ converges in the norm (1.75), which means that $C^{0,\alpha}(\partial S)$ is complete.

1.69. Definition. Let X and Y be normed spaces. A linear operator $K : X \to Y$ is called *compact* if it maps any bounded set in X into a relatively compact set in Y (that is, a set in which every sequence contains a convergent subsequence).

1.70. Theorem. *If $k(z, \zeta)$ is a proper γ-singular kernel on ∂S, $\gamma \in [0, 1)$, then the operator K defined by*

$$(K\varphi)(z) = \int_{\partial S} k(z, \zeta)\varphi(\zeta)\, d\zeta, \quad z \in \partial S, \tag{1.78}$$

is a compact operator from $C^{0,\alpha}(\partial S)$ into $C^{0,\alpha}(\partial S)$, with $\alpha = 1 - \gamma$ for $\gamma \in (0,1)$ and any $\alpha \in (0,1)$ for $\gamma = 0$.

Proof. According to Theorem 1.30 and the fact that $C^{0,\alpha}(\partial S) \subset C(\partial S)$, the operator $K : C^{0,\alpha}(\partial S) \to C^{0,\alpha}(\partial S)$ is well defined.

Let $M_1 \subset C^{0,\alpha}(\partial S)$ be a bounded set, that is,

$$\|\varphi\|_\alpha \leq c = \text{const} > 0 \quad \text{for all} \quad \varphi \in M_1. \tag{1.79}$$

Also, let $\{\theta_n\}_{n=1}^\infty \subset M_2 = K(M_1)$. We denote by $\{\varphi_n\}_{n=1}^\infty$ a sequence in M_1 such that $\theta_n = K\varphi_n$, $n = 1, 2, \ldots$

In view of (1.75), the inequality (1.79) implies that

$$\sup_{z \in \partial S} |\varphi_n(z)| \leq c,$$

$$|\varphi_n(z) - \varphi_n(z')| \leq c|z - z'|^\alpha$$

for all $n = 1, 2, \ldots$ and all x, $x' \in \partial S$; in other words, $\{\varphi_n\}_{n=1}^\infty$ is uniformly bounded and equicontinuous in $C(\partial S)$. By the Arzelà-Ascoli theorem [4], it contains a uniformly convergent subsequence. For simplicity, we denote this subsequence again by $\{\varphi_n\}_{n=1}^\infty$. Hence, there is a $\varphi \in C(\partial S)$ such that

$$\|\varphi_n - \varphi\|_\infty \to 0 \quad \text{as} \quad n \to \infty. \tag{1.80}$$

Let $\theta = K\varphi$. By Theorem 1.30, $\theta \in C^{0,\alpha}(\partial S)$. For $z \in \partial S$ we have

$$|\theta_n(z) - \theta(z)| \leq \int_{\partial S} |k(z, \zeta)| |\varphi_n(\zeta) - \varphi(\zeta)| d\zeta$$

$$\leq c_1 \sup_{x \in \partial S} |\varphi_n(z) - \varphi(z)| \int_{\partial S} |z - \zeta|^{-\gamma} d\zeta,$$

consequently, by Theorem 1.29,

$$\|\theta_n - \theta\|_\infty \leq c_2 \|\varphi_n - \varphi\|_\infty, \quad n = 1, 2, \ldots,$$

where c_1 and c_2 are positive constants. On the other hand, by Theorem 1.30,
$$|\theta_n - \theta|_\alpha \leq c_3 \|\varphi_n - \varphi\|_\alpha, \quad n = 1, 2, \ldots$$
The last two inequalities, (1.75) and (1.80) yield
$$\|\theta_n - \theta\|_\alpha \to \infty \quad \text{as} \quad n \to \infty,$$
which proves that K is compact on $C^{0,\alpha}(\partial S)$.

1.71. Definition. Let X and Y be two vector spaces over $\mathbf{C}$. A mapping $(\cdot, \cdot) : X \times Y \to \mathbf{C}$ is called a *non-degenerate bilinear form* if

(i) for any $\varphi \in X$, $\varphi \neq 0$, there is a $\psi \in Y$ such that $(\varphi, \psi) \neq 0$, and for any $\psi \in Y$, $\psi \neq 0$, there is a $\varphi \in X$ such that $(\varphi, \psi) \neq 0$;

(ii) for any $\varphi_1, \varphi_2, \varphi \in X$, $\psi_1, \psi_2, \psi \in Y$ and $\alpha_1, \alpha_2 \in \mathbf{C}$
$$(\alpha_1 \varphi_1 + \alpha_2 \varphi_2, \psi) = \alpha_1(\varphi_1, \psi) + \alpha_2(\varphi_2, \psi),$$
$$(\varphi, \alpha_1 \psi_1 + \alpha_2 \psi_2) = \alpha_1(\varphi, \psi_1) + \alpha_2(\varphi, \psi_2).$$

1.72. Definition. By a *dual system* (X, Y) we understand a pair of normed spaces X and Y together with a non-degenerate bilinear form $(\cdot, \cdot) : X \times Y \to \mathbf{C}$.

1.73. Definition. Let (X, Y) be a dual system with bilinear form $(\cdot, \cdot)$. Two operators $K : X \to X$ and $K^* : Y \to Y$ are called *adjoint* if
$$(K\varphi, \psi) = (\varphi, K^*\psi) \tag{1.81}$$
for all $\varphi \in X$ and $\psi \in Y$.

1.74. Remark. It can be shown without difficulty [4] that if an operator $K : X \to X$ has an adjoint $K^* : Y \to Y$ in a dual system (X, Y), then K^* is unique, and both K and K^* are linear.

1.75. Definition. Let (X, Y) be a dual system with bilinear form $(\cdot, \cdot)$, $K : X \to X$ an operator that has a (unique) adjoint $K^* : Y \to Y$, I the identity operator (which, for simplicity, is denoted by the same symbol regardless of the space where it acts), and $w \in \mathbb{C}$, $w \neq 0$, and consider the equations

$$(K - wI)\varphi = f, \quad f \in X, \tag{K}$$
$$(K^* - wI)\psi = g, \quad g \in Y, \tag{K*}$$

together with their homogeneous versions (K_0) and (K_0^*). We say that the Fredholm Alternative holds for K in (X, Y) if either

(i) (K_0) has only the zero solution, in which case so does (K_0^*), and (K) and (K*) have unique solutions for any $f \in X$ ad $g \in Y$, respectively, or

(ii) (K_0) and (K_0^*) have the same finite number of linearly independent solutions $\{\varphi_1, \ldots, \varphi_n\}$ and $\{\psi_1, \ldots, \psi_n\}$, and (K) and (K*) are soluble respectively if and only if

$$(f, \psi_i) = 0, \quad (g, \varphi_i) = 0, \quad i = 1, \ldots, n.$$

1.76. Theorem. Let (X, Y) be a dual system and $K : X \to X$ a compact linear operator that has a (unique) compact adjoint $K^* : Y \to Y$. Then the Fredholm Alternative holds for K in (X, Y).

A full, detailed proof of this assertion can be found, for example, in the monograph [4].

1.77. Remark. Let K be the operator defined by (1.78), and consider the dual system $(C^{0,\alpha}(\partial S), C^{0,\alpha}(\partial S))$, $\alpha \in (0, 1)$, with the bilinear form

$$(\varphi, \psi) = \int_{\partial S} \varphi(\zeta)\psi(\zeta)\, d\zeta, \quad \varphi, \psi \in C^{0,\alpha}(\partial S), \tag{1.82}$$

which is easily seen to satisfy the conditions in Definition 1.71. From (1.81) and (1.82) we have

$$(K\varphi, \psi) = \int_{\partial S}\left[\int_{\partial S} k(z, \zeta)\varphi(\zeta)\,d\zeta\right]\psi(z)\,dz$$

$$= \int_{\partial S}\varphi(\zeta)\left[\int_{\partial S} k(z, \zeta)\psi(z)\,dz\right]d\zeta = (\varphi, K^*\psi),$$

where

$$(K^*\varphi)(z) = \int_{\partial S} k^*(z, \zeta)\varphi(\zeta)\,d\zeta,$$

with $k^*(z, \zeta) = k(\zeta, z)$.

This means that if $k(z, \zeta)$ is a proper $(1-\alpha)$-singular kernel on ∂S with respect to both z and ζ, then, by Theorems 1.70 and 1.76, the Fredholm Alternative holds for K.

The Fredholm Alterative does not hold in general for operators with 1-singular kernels. However, there is a class of such operators for which the assertion remains true.

Theorem 1.67 enables us to introduce the following concept.

1.78. Definition. An operator $K : C^{0,\alpha}(\partial S) \to C^{0,\alpha}(\partial S)$, $\alpha \in (0, 1)$, is called α-*regular singular* if it is defined by an expression of the form

$$(K\varphi)(z) = \int_{\partial S} \frac{\hat{k}(z, \zeta)}{\zeta - z}\varphi(\zeta)\,d\zeta, \quad z \in \partial S, \tag{1.83}$$

where $\hat{k}(z, \zeta)$ belongs to $C^{0,\alpha}(\partial S)$ with respect to each variable, uniformly relative to the ther one, and satisfies the inequality

$$|\hat{k}(z, \zeta) - \hat{k}(z', \zeta)| \leq c|z - z'||z - \zeta|^{\alpha-1}, \quad c = \text{const} > 0,$$

for all $z, z', \zeta \in \partial S$ such that $0 < |z-z'| < \frac{1}{2}|z-\zeta|$. (The value of $\hat{k}(z, \zeta)$ at $z = \zeta$ may also be understood in the sense of continuous extension.)

1.79. Theorem. *If $k(x, y)$ is a proper γ-singular kernel on ∂S, $\gamma \in [0, 1)$, with respect to both x and y, then the operator K on $C^{0,1-\gamma}(\partial S)$ defined by*

$$(K\varphi)(x) = \int_{\partial S} k(x, y)\varphi(y)\,ds(y), \quad x \in \partial S,$$

is $(1-\gamma)$-regular singular, and $\hat{k}(z, \zeta)$ in (1.83) satisfies

$$\hat{k}(z, z) = 0, \quad z \in \partial S.$$

Proof. Clearly, $(K\varphi)(x)$ is an improper integral for all $x \in \partial S$.

In accordance with our notational convention, we write

$$\int_{\partial S} k(x, y)\varphi(y)\widetilde{ds(y)} = \int_{\partial S} k(z, \zeta)\varphi(\zeta)\bar{\vartheta}(\zeta)\,d\zeta = \int_{\partial S} \frac{\hat{k}(z, \zeta)}{\zeta - z}\,d\zeta,$$

where $\vartheta(z)$ is defined by (1.68) and

$$\hat{k}(z, \zeta) = (\zeta - z)k(z, \zeta)\bar{\vartheta}(\zeta).$$

By Definition 1.23, $\hat{k}(z, z) = 0$ in the sense of continuous extension.

Let $z, z', \zeta \in \partial S$ be such that $0 < |z - z'| < \frac{1}{2}|z - \zeta|$. In this case

$$|z' - \zeta| \geq |z - \zeta| - |z - z'| > |z - \zeta| - \tfrac{1}{2}|z - \zeta| = \tfrac{1}{2}|z - \zeta|. \tag{1.84}$$

Since $|\vartheta(\zeta)| = 1$, $\zeta \in \partial S$, we have

$$|\hat{k}(z, \zeta) - \hat{k}(z', \zeta)| = |(\zeta - z)k(z, \zeta) - (\zeta - z')k(z', \zeta)|$$
$$\leq |z - \zeta||k(z, \zeta) - k(z', \zeta)| + |z - z'||k(z', \zeta)|$$
$$\leq c_1|z - z'|(|z - \zeta|^{-\gamma} + |z' - \zeta|^{-\gamma}),$$

which, on the basis of (1.84), shows that

$$|\hat{k}(z, \zeta) - \hat{k}(z', \zeta)| \le c_2 |z - z'| |z - \zeta|^{-\gamma}$$

and that

$$|\hat{k}(z, \zeta) - \hat{k}(z', \zeta)| \le c_3 |z - z'|^{1-\gamma},$$

where the constants $c_1, c_2, c_3 > 0$ are independent of z, z' and ζ.
If $|z - z'| \ge \frac{1}{2}|z - \zeta|$, then

$$|z' - \zeta| \le |z - z'| + |z - \zeta| \le 3|z - z'|,$$

consequently,

$$|\hat{k}(z, \zeta) - \hat{k}(z', \zeta)| \le |z - \zeta| |k(z, \zeta)| + |z' - \zeta| |k(z', \zeta)|$$
$$\le c_1(|z - \zeta|^{1-\gamma} + |z' - \zeta|^{1-\gamma}) \le c_4 |z - z'|^{1-\gamma},$$

where c_4 is independent of z, z' and ζ.

The Hölder continuity of $\hat{k}(z, \zeta)$ with respect to ζ is proved similarly by writing

$$|\hat{k}(z, \zeta) - \hat{k}(z, \zeta')| \le |\zeta - z| |k(z, \zeta)| |\vartheta(\zeta) - \vartheta(\zeta')|$$
$$+ |\zeta - z| |k(z, \zeta) - k(z, \zeta')| + |\zeta - \zeta'| |k(z, \zeta')|$$

and using the fact that $\vartheta \in C^{0,\alpha}(\partial S)$.

1.80. Definition. Consider an equation of the form

$$(K - \omega I)\varphi = f \quad \text{on} \quad \partial S, \tag{1.85}$$

where K is an α-regular singular operator, φ and f are (3×1)-matrices in $C^{0,\alpha}(\partial S)$, $\alpha \in (0, 1)$, $\omega \in \mathbf{C}$, $\omega \ne 0$, and $\det\left[-\omega E \pm \pi i \hat{k}(z, z)\right]$ (see

Definition 1.78) do not vanish on ∂S. The number [27]

$$\varrho = \frac{1}{2\pi}\left[\arg\frac{\det\left(-\omega E - \pi i \hat{k}(z, z)\right)}{\det\left(-\omega E + \pi i \hat{k}(z, z)\right)}\right]_{\partial S}, \qquad (1.86)$$

where $[\theta(z)]_{\partial S}$ denotes the change in $\theta(z)$ as z traverses ∂S once anti-clockwise, is called the *index* of the equation (1.85).

When (1.85) is a scalar equation, the symbol det is dropped in (1.86).

1.81. Remark. Let K be an α-regular singular operator, $\alpha \in (0,1)$, and consider the dual system $(C^{0,\alpha}(\partial S), C^{0,\alpha}(\partial S))$ with the bilinear form (1.82). Then from (1.81) and (1.83) we conclude that the kernel of the adjoint K^* of K is $-\hat{k}(\zeta, z)/(\zeta - z)$, which implies that, by (1.86), the index of the equation

$$(K^* - \omega I)\psi = g \quad \text{on} \quad \partial S, \quad g \in C^{0,\alpha}(\partial S),$$

is equal to $-\varrho$.

1.82. Theorem. *If K is an α-regular singular operator, $\alpha \in (0,1)$, such that the index of the equation (K) is zero, then the Fredholm Alternative holds for K in the dual system $(C^{0,\alpha}(\partial S), C^{0,\alpha}(\partial S))$ with the bilinear form (1.82).*

A comprehensive discussion of this assertion can be found in [27] and [24]. Its proof consists of two stages. First, it is shown that we can always find an α-regular singular operator L and a $\vartheta \in \mathbf{C}$, $\vartheta \neq 0$, such that the equation

$$(L - \vartheta I)(K - \omega I)\varphi = (L - \vartheta I)f$$

is of the form

$$(\tilde{K} - \tilde{\omega}I)\tilde{\varphi} = \tilde{f}, \qquad (\tilde{K})$$

where $\tilde{\omega} \in \mathbf{C}$, $\tilde{\omega} \neq 0$, $\tilde{f} \in C^{0,\alpha}(\partial S)$, $\tilde{K}$ is an integral operator defined by

$$(\tilde{K}\tilde{\varphi})(z) = \int_{\partial S} \tilde{k}(z, \zeta)\tilde{\varphi}(\zeta)\, d\zeta, \quad z \in \partial S,$$

and $\tilde{k}(z, \zeta)$ is a proper $(1 - \alpha)$-singular kernel on ∂S with respect to both z and ζ. By Remark 1.77, the Fredholm Alternative holds for the operator K in the dual system $(C^{0,\alpha}(\partial S), C^{0,\alpha}(\partial S))$ with the bilinear form (1.82). The second part of the proof consists in showing that, since the indices of both (K) and (K*) are zero (the latter according to Remark 1.81), (K), ($\tilde{K}$) and (K*), ($\tilde{K}$*) have respectively the same solutions.

1.83. Remark. Let $K : C^{0,\alpha}(\partial S) \to C^{0,\alpha}(\partial S)$, $\alpha \in (0,1)$, be an operator of the form

$$(K\varphi)(z) = \int_{\partial S} k(z, \zeta)\varphi(\zeta)\, d\zeta, \quad z \in \partial S,$$

and consider the corresponding equation (K)

$$-\omega\varphi(z) + \int_{\partial S} k(z, \zeta)\varphi(\zeta)\, d\zeta = f(z), \quad z \in \partial S.$$

Suppose now that, for f and ω real, when we change from z and ζ to x and y the transformed equation

$$-\omega\varphi(x) + \int_{\partial S} k(x, y)\vartheta(y)\varphi(y)\, ds(y) = f(x), \quad x \in \partial S,$$

where ϑ is defined by (1.68), is real. By Remark 1.77, the kernel of the adjoint operator K_c^* in the complex dual system $(C^{0,\alpha}(\partial S), C^{0,\alpha}(\partial S))$ with the bilinear form (1.82) is

$$k_c^*(z, \zeta) = k(\zeta, z) = k(y, x)\vartheta(y).$$

On the other hand, it is easy to see that the kernel of the adjoint K_r^* in the real dual system $(C^{0,\alpha}(\partial S), C^{0,\alpha}(\partial S))$ with the bilinear form

$$(\varphi, \psi) = \int_{\partial S} \varphi(y)\psi(y)\, ds(y) \qquad (1.87)$$

is

$$k_r^*(x, y) = k(y, x)\vartheta(x),$$

which is different from $k_c^*(z, \zeta)$. In [27] it is shown that if the Fredholm Alternative holds for the operator K in the complex system, then it also holds for it in the real system, provided that we restrict ourselves to real solutions of (K) and (K*) with $K^* = K_r^*$.

1.84. Remark. If $C^{0,\alpha}(\partial S)$ is understood as a space of (3×1)-matrix functions, then φ is replaced by φ^T in (1.82) and (1.87).

2 Bending of elastic plates

2.1. The two-dimensional plate model

We consider the averaging operators $\mathcal{I}_{\alpha-1}$ and $\mathcal{J}_{\alpha-1}$ defined by

$$(\mathcal{I}_{\alpha-1}g)(x_\gamma) = \frac{1}{h_0}\left[x_3^{\alpha-1}g(x_i)\right]_{x_3=-h_0/2}^{x_3=h_0/2},$$

$$(\mathcal{J}_{\alpha-1}g)(x_\gamma) = \frac{1}{h_0}\int_{-h_0/2}^{h_0/2} x_3^{\alpha-1}g(x_i)\,dx_3. \tag{2.1}$$

Setting

$$\begin{aligned} N_{\alpha\beta} &= \mathcal{J}_1 t_{\alpha\beta}, \\ N_{3\alpha} &= \mathcal{J}_0 t_{3\alpha}, \\ F_\alpha &= -(\mathcal{J}_1 f_\alpha + \mathcal{I}_1 t_{3\alpha}), \\ F_3 &= -(\mathcal{J}_0 f_3 + \mathcal{I}_0 t_{33}), \end{aligned} \tag{2.2}$$

from (1.1) and (2.1) we obtain the equilibrium equations

$$\begin{aligned} N_{\alpha\beta,\beta} - N_{3\alpha} &= F_\alpha, \\ N_{3\beta,\beta} &= F_3. \end{aligned} \tag{2.3}$$

By analogy with (1.3), for a direction $n = (n_1, n_2)^\mathrm{T}$ in the middle plane we write

$$\begin{aligned} N_\alpha &= N_{\alpha\beta} n_\beta, \\ N_3 &= N_{3\beta} n_\beta. \end{aligned} \tag{2.4}$$

2.1. Remark. It is easy to verify that $N_{3\alpha}$, $N_{\alpha\alpha}$ (α not summed) and $N_{12} = N_{21}$ are respectively the averaged transverse shear forces and averaged bending and twisting moments with respect to the middle plane, acting on the face of a vertical cross-section element of the plate perpendicular to the x_α-axis. Similarly, $\mathcal{J}_0 f_3$ and $\mathcal{J}_1 f_\alpha$ are the averaged body force and moments, while $\mathcal{I}_0 t_{33}$ and $\mathcal{I}_1 t_{3\alpha}$ are the resultant averaged force and moments acting on the faces $x_3 = \pm\tfrac{1}{2} h_0$. For the sake of simplicity, from now on we will omit the word 'averaged' when referring to forces and moments in the plate.

It can also be seen that the components of the moment with respect to the middle plane and the transverse shear force in a direction n are $\varepsilon_{\beta\alpha} N_\beta$ and N_3, respectively. If the moment is computed with respect to the origin of coordinates, then its components are $\varepsilon_{\beta\alpha}(N_\beta - x_\beta N_3)$. Clearly, knowing the N_i at a point is equivalent to knowing the $\varepsilon_{\beta\alpha}(N_\beta - x_\beta N_3)$ and N_3.

To avoid a clash of notation with the harmonic single layer potential, in what follows we write u_i in place of v_i on the right-hand side of (1.6). Thus, from (1.6), (1.2) and (2.1) we obtain the constitutive relations

$$N_{\alpha\beta} = h^2 \big[\lambda u_{\gamma,\gamma}\delta_{\alpha\beta} + \mu(u_{\alpha,\beta} + u_{\beta,\alpha})\big],$$
$$N_{3\alpha} = \mu(u_\alpha + u_{3,\alpha}), \qquad (2.5)$$

where $h^2 = \tfrac{1}{12} h_0^2$. The same formulae and (1.3) show that

$$N_\alpha = \mathcal{I}_1 t_\alpha, \quad N_3 = \mathcal{I}_0 t_3.$$

To establish the compatibility conditions for the $N_{i\alpha}$, first we use (2.5) to deduce that

$$u_{1,1} = \frac{1}{2h^2 \mu}\big[(1-\sigma)N_{11} - \sigma N_{22}\big],$$
$$u_{2,2} = \frac{1}{2h^2 \mu}\big[(1-\sigma)N_{22} - \sigma N_{11}\big],$$
$$u_{1,2} + u_{2,1} = \frac{1}{h^2 \mu} N_{12}, \qquad (2.6)$$
$$u_{1,2} - u_{2,1} = \frac{1}{\mu}(N_{31,2} - N_{32,1}),$$

where
$$\sigma = \frac{\lambda}{2(\lambda + \mu)}$$
is Poisson's ratio. The compatibility conditions are then derived by equating $u_{\alpha,12}$ with $u_{\alpha,21}$ after computing these derivatives from (2.6). As a result we obtain

$$h^2(N_{31,12} - N_{32,11}) + N_{12,1} - (1-\sigma)N_{11,2} + \sigma N_{22,2} = 0,$$
$$h^2(N_{32,12} - N_{31,22}) + N_{12,2} + \sigma N_{11,1} - (1-\sigma)N_{22,1} = 0. \qquad (2.7)$$

From (2.3) and (2.5) we find that the equilibrium equations in terms of the displacements are

$$A(\partial_x)u(x) = F(x), \qquad (2.8)$$

where $A(\partial_x) = A(\partial/\partial x_\gamma)$ and $A(\xi) = A(\xi_\gamma)$ is the matrix

$$\begin{pmatrix} h^2\mu\Delta + h^2(\lambda+\mu)\xi_1^2 - \mu & h^2(\lambda+\mu)\xi_1\xi_2 & -\mu\xi_1 \\ h^2(\lambda+\mu)\xi_1\xi_2 & h^2\mu\Delta + h^2(\lambda+\mu)\xi_2^2 - \mu & -\mu\xi_2 \\ \mu\xi_1 & \mu\xi_2 & \mu\Delta \end{pmatrix}, \qquad (2.9)$$

$u = (u_1, u_2, u_3)^T$, $F = (F_1, F_2, F_3)^T$, and $\Delta = \xi_\alpha\xi_\alpha$. Then N can be written as

$$N = T(\partial_x; n)u(x), \qquad (2.10)$$

where $T(\partial_x; n) = T(\partial/\partial x_\gamma; n)$ and $T(\xi; n) = T(\xi_\gamma; n_\delta)$ is the matrix

$$\begin{pmatrix} h^2(\lambda+2\mu)n_1\xi_1 + h^2\mu n_2\xi_2 & h^2\mu n_2\xi_1 + h^2\lambda n_1\xi_2 & 0 \\ h^2\lambda n_2\xi_1 + h^2\mu n_1\xi_2 & h^2\mu n_1\xi_1 + h^2(\lambda+2\mu)n_2\xi_2 & 0 \\ \mu n_1 & \mu n_2 & \mu n_\alpha\xi_\alpha \end{pmatrix}.$$
$$(2.11)$$

For brevity we also write $T(\partial_x; \nu) \equiv T(\partial_x) \equiv T$.

75

From (1.4), (1.6), (2.1), and (2.2) we see that the internal energy density per unit area of the middle plane is

$$E(u, u) = \mathcal{J}_0 \mathcal{E} = \tfrac{1}{4} N_{\alpha\beta}(u_{\alpha,\beta} + u_{\beta,\alpha}) + \tfrac{1}{2} N_{3\alpha}(u_\alpha + u_{3,\alpha})$$
$$= \tfrac{1}{2}\{h^2 [\lambda u_{\alpha,\alpha} u_{\beta,\beta} + \mu u_{\alpha,\beta}(u_{\alpha,\beta} + u_{\beta,\alpha})]$$
$$+ \mu(u_\alpha + u_{3,\alpha})(u_\alpha + u_{3,\alpha})\}. \quad (2.12)$$

Throughout what follows we assume that the Lamé constants satisfy the conditions

$$\lambda + \mu > 0, \quad \mu > 0. \quad (2.13)$$

2.2. Theorem. $E(u, u)$ *is a positive quadratic form and* (2.8) *an elliptic system.*

Proof. From (2.12) it follows that

$$E(u, u) = \tfrac{1}{2}\{h^2 [E_0(u, u) + \mu(u_{1,2} + u_{2,1})^2]$$
$$+ \mu[(u_1 + u_{3,1})^2 + (u_2 + u_{3,2})^2]\}, \quad (2.14)$$

where

$$E_0(u, u) = (\lambda + 2\mu)u_{1,1}^2 + 2u_{1,1}u_{2,2} + (\lambda + 2\mu)u_{2,2}^2. \quad (2.15)$$

We now easily verify that (2.10) are necessary and sufficient conditions for $E_0(u, u)$ to be a positive definite quadratic form.

The second part of the assertion is obtained from the fact that, by (2.9), the matrix $A_0(\xi)$ corresponding to the second order derivatives in the system (2.8) is invertible for all $\xi \neq 0$ since

$$\det A_0(\xi) = a_1(\xi_1^2 + \xi_2^2)^3,$$

where

$$a_1 = h^4 \mu^2 (\lambda + 2\mu) > 0.$$

2.3. Theorem. $E(u, u) = 0$ if and only if

$$u(x) = (c_1, c_2, c_0 - c_1 x_1 - c_2 x_2)^\mathrm{T}, \qquad (2.16)$$

where $c_0, c_\alpha = \mathrm{const}$.

Proof. Replacing (2.16) in (2.12), we see immediately that $E(u, u) = 0$. Conversely, suppose that $E(u, u) = 0$. From (2.14) and (2.15) we get

$$u_{1,1} = u_{2,2} = 0,$$
$$u_{1,2} + u_{2,1} = 0,$$
$$u_{3,\alpha} + u_\alpha = 0.$$

The first two equations yield the equalities

$$u_1 = f_1(x_2),$$
$$u_2 = f_2(x_1),$$

which, replaced in the second relation, lead to

$$f_1(x_2) = kx_2 + c_1,$$
$$f_2(x_1) = -kx_1 + c_2,$$

where k, c_1 and c_2 are arbitrary constants. Using the compatibility condition for u_3, that is, $u_{3,12} = u_{3,21}$, from the last two equations we find that $k = 0$. Hence, $u_\alpha = c_\alpha$, so that

$$u_{3,\alpha} = -c_\alpha.$$

Integrating the equation for $\alpha = 1$ and substituting the result into that for $\alpha = 2$, we obtain

$$u_3 = c_0 - c_\alpha x_\alpha,$$

where c_0 is an arbitrary constant.

2.4. Remark. Since the three-dimensional displacement field we are investigating is of the form

$$\bigl(x_3 u_1(x_1, x_2),\ x_3 u_2(x_1, x_2),\ u_3(x_1, x_2)\bigr)^{\mathrm{T}},$$

the most general admissible translation and rotation vectors are respectively of the form

$$(0,\, 0,\, a)^{\mathrm{T}} \quad \text{and} \quad (-x_3 b_2,\ x_3 b_1,\ x_1 b_2 - x_2 b_1)^{\mathrm{T}},$$

where a, b_1 and b_2 are arbitrary constants. Therefore, setting $a = c_0$, $b_1 = c_2$ and $b_2 = -c_1$, we conclude that (2.16) represents an arbitrary rigid displacement.

2.5. Theorem. *If* $u \in C^2(S^+) \cap C^1(\bar{S}^+)$, *then*

$$\int_{S^+} (A_{\alpha i} - x_\alpha A_{3i}) u_i\, da = \int_{\partial S} (T_{\alpha i} - x_\alpha T_{3i}) u_i\, ds,$$

$$\int_{S^+} A_{3i} u_i\, da = \int_{\partial S} T_{3i} u_i\, ds.$$

Proof. Since (2.3) and (2.8) are the same system and (2.4) and (2.10) the same (3×1)-matrix, by the Divergence Theorem,

$$\int_{S^+} (A_{\alpha i} - x_\alpha A_{3i}) u_i\, da = \int_{S^+} (N_{\alpha\beta,\beta} - N_{3\alpha} - x_\alpha N_{3\beta,\beta})\, da$$

$$= \int_{\partial S} (N_{\alpha\beta} - x_\alpha N_{3\beta}) \nu_\beta\, ds = \int_{\partial S} (T_{\alpha i} - x_\alpha T_{3i}) u_i\, ds,$$

$$\int_{S^+} A_{3i} u_i\, da = \int_{S^+} N_{3\beta,\beta}\, da = \int_{\partial S} N_{3\beta} \nu_\beta\, ds = \int_{\partial S} T_{3i} u_i\, ds.$$

2.6. Theorem (Betti formula). *If $u \in C^2(S^+) \cap C^1(\bar{S}^+)$ is a solution of the homogeneous system (2.8), then*

$$2 \int_{S^+} E(u, u)\, da = \int_{\partial S} u^T Tu\, ds.$$

Proof. Using (2.8), (2.3), (2.10), (2.4), the Divergence Theorem, and (2.12), we find that for any $u \in C^2(S^+) \cap C^1(\bar{S}^+)$

$$0 = \int_{S^+} u^T Au\, da = \int_{S^+} \left[(N_{\alpha\beta,\beta} - N_{3\alpha})u_\alpha + N_{3\alpha,\alpha} u_3\right] da$$

$$= -\int_{S^+} \left[N_{\alpha\beta} u_{\alpha,\beta} + N_{3\alpha}(u_\alpha + u_{3,\alpha})\right] da + \int_{\partial S} N_i u_i\, ds$$

$$= -2 \int_{S^+} E(u, u)\, da + \int_{\partial S} u^T Tu\, ds,$$

which yields the result.

2.7. Theorem (reciprocity relation). *If $u, \tilde{u} \in C^2(S^+) \cap C^1(\bar{S}^+)$, then*

$$\int_{S^+} (\tilde{u}^T Au - u^T A\tilde{u})\, da = \int_{\partial S} (\tilde{u}^T Tu - u^T T\tilde{u})\, ds.$$

Proof. Let $N_{i\beta}$ and $\tilde{N}_{i\beta}$ be the moments and transverse shear forces generated by the displacements u and $\tilde{u}$, respectively. Using the equivalence of (2.3), (2.8) and (2.4), (2.10) and the Divergence Theorem, we find that

$$\int_{S^+} (\tilde{u}^T Au - u^T A\tilde{u})\, da$$

$$= \int_{S^+} \left[(N_{\alpha\beta,\beta} - N_{3,\alpha})\tilde{u}_\alpha + N_{3\alpha,\alpha}\tilde{u}_3 - (\tilde{N}_{\alpha\beta,\beta} - \tilde{N}_{3\alpha})u_\alpha - \tilde{N}_{3\alpha,\alpha} u_3\right] da$$

$$= \int_{\partial S} (N_i \tilde{u}_i - \tilde{N}_i u_i) \, ds$$

$$- \int_{S^+} [N_{\alpha\beta} \tilde{u}_{\alpha,\beta} + N_{3\alpha}(\tilde{u}_\alpha + \tilde{u}_{3,\alpha}) - \tilde{N}_{\alpha\beta} u_{\alpha,\beta} - \tilde{N}_{3\alpha}(u_\alpha + u_{3,\alpha})] \, da$$

$$= \int_{\partial S} (\tilde{u}^T T u - u^T T \tilde{u}) \, ds,$$

since, by (2.5), the integrand of the second integral vanishes in S^+.

2.2. Singular solutions

We seek a Galerkin representation for the solution of (2.8). Following the method described in [6], if $A^*(\xi)$ is the adjoint of $A(\xi)$, then

$$u(x) = A^*(\partial_x) B(x), \tag{2.17}$$

where B is the solution of

$$(\det A)(\partial_x) B(x) = F(x). \tag{2.18}$$

More explicitly, from (2.9) we find that

$$\begin{aligned}
A^*_{\alpha\beta}(\xi) &= h^2 \mu(\lambda + 2\mu) \delta_{\alpha\beta} \Delta\Delta - h^2 \mu(\lambda + \mu) \Delta \xi_\alpha \xi_\beta - \mu^2 \xi_\alpha \xi_\beta, \\
A^*_{33}(\xi) &= h^4 \mu(\lambda + 2\mu) \Delta\Delta - h^2 \mu(\lambda + 3\mu) \Delta + \mu^2, \\
A^*_{\alpha 3}(\xi) &= -A^*_{3\alpha}(\xi) = \mu^2 \xi_\alpha (h^2 \Delta - 1),
\end{aligned} \tag{2.19}$$

and

$$\det A(\xi) = a_1 \Delta\Delta(\Delta - h^{-2}). \tag{2.20}$$

Taking in turn each component of F equal to $-\delta(|x - y|)$, where δ is the Dirac distribution, and setting the other two equal to zero, from (2.17) and (2.18) we obtain the matrix of fundamental solutions

$$D(x, y) = A^*(\partial_x)t(x, y), \qquad (2.21)$$

where, by (2.18) and (2.20), $t(x, y)$ is a solution of

$$\Delta\Delta(\Delta - h^{-2})t(x, y) = -a_1^{-1}\delta(|x - y|). \qquad (2.22)$$

We seek t of the form

$$t(x, y) = b_1 \ln |x - y| + b_2 |x - y|^2 \ln |x - y| + b_3 K_0(h^{-1}|x - y|),$$

where K_0 is the modified Bessel function of order zero. Replacing this in (2.22) and taking into account the fact that, with respect to x,

$$\Delta(\ln |x - y|) = 2\pi\delta(|x - y|),$$
$$\Delta\Delta(|x - y|^2 \ln |x - y|) = 8\pi\delta(|x - y|),$$
$$(\Delta - h^{-2})K_0(h^{-1}|x - y|) = -2\pi\delta(|x - y|),$$

we deduce that

$$t(x, y) = t(|x - y|)$$
$$= a_2\big[(4h^2 + |x - y|^2)\ln|x - y| + 4h^2 K_0(h^{-1}|x - y|)\big], \qquad (2.23)$$

where

$$a_2 = \frac{1}{8\pi h^2 \mu^2(\lambda + 2\mu)}.$$

In view of (2.21)–(2.23),

$$D(x, y) = (D(y, x))^{\mathrm{T}}. \qquad (2.24)$$

Along with $D(x, y)$ we consider the matrix of singular solutions

$$P(x, y; n) = (T(\partial_y; n)D(y, x))^{\mathrm{T}}, \qquad (2.25)$$

writing, for simplicity, $P(x, y; \nu(y)) \equiv P(x, y)$.

To determine the behaviour of $D(x, y)$ and $P(x, y)$ in the neighbourhood of $x = y$, we note [1] that, as $\xi \to 0$,

$$K_0(\xi) = -(1 + \tfrac{1}{4}\xi^2 + \tfrac{1}{64}\xi^4 + \cdots) \ln \xi,$$

so from (2.23) we deduce that for $|x - y|$ small

$$t(x, y) = a_3 |x - y|^4 \ln |x - y| + \tilde{t}(x, y), \qquad (2.26)$$

where $\tilde{t} \in C^5(\mathbf{R}^2)$ and

$$a_3 = -\frac{1}{128\pi h^4 \mu^2 (\lambda + 2\mu)}.$$

We denote by $\{E_{ij}\}$ the standard ordered basis for the vector space of (3×3)-matrices. From (2.21), (2.19), (2.23), (2.25), (2.11), and (2.26) we now find that for y close to x

$$D(x, y) = -\frac{1}{2\pi} \ln |x - y| \left(a_4 E_{\gamma\gamma} + \frac{1}{\mu} E_{33} \right)$$
$$- 2a_2 \mu^2 \frac{(x_\alpha - y_\alpha)(x_\beta - y_\beta)}{|x - y|^2} E_{\alpha\beta} + \tilde{D}(x, y), \qquad (2.27)$$

$$P(x, y) = -\frac{1}{2\pi} \left\{ \mu' \varepsilon_{\alpha\beta} \left[\frac{\partial}{\partial s(y)} \ln |x - y| \right] E_{\alpha\beta} + \left[\frac{\partial}{\partial \nu(y)} \ln |x - y| \right] E \right.$$
$$- (\lambda' + \mu') \varepsilon_{\alpha\gamma} \left[\frac{\partial}{\partial s(y)} \frac{(x_\alpha - y_\alpha)(x_\beta - y_\beta)}{|x - y|^2} \right] E_{\gamma\beta}$$
$$+ \tfrac{1}{2} \varepsilon_{\alpha\beta} \left[\frac{\partial}{\partial s(y)} ((x_\alpha - y_\alpha) \ln |x - y|) \right] \left(\lambda' E_{3\beta} + \frac{1}{h^2} E_{\beta 3} \right)$$
$$\left. - \tfrac{1}{2} \left[\frac{\partial}{\partial \nu(y)} ((x_\alpha - y_\alpha) \ln |x - y|) \right] E_{3\alpha} \right\}$$
$$+ \tilde{P}(x, y), \qquad (2.28)$$

where E is the identity (3×3)-matrix, $\tilde{D}(x, y)$ and $\tilde{P}(x, y)$ satisfy the conditions of Theorem 1.53 (or 1.54) with any $\gamma \in (0, 1)$,

$$a_4 = \frac{\lambda + 3\mu}{2h^2 \mu(\lambda + 2\mu)},$$

and

$$\lambda' = \frac{\lambda}{\lambda + 2\mu}, \quad \mu' = \frac{\mu}{\lambda + 2\mu}. \tag{2.29}$$

2.8. Theorem. *The columns of $D(x, y)$ and $P(x, y; n)$ are solutions of (2.8) at all $x \in \mathbf{R}^2$, $x \neq y$, and for any direction n independent of x.*

Proof. Since $A(\xi)A^*(\xi) = (\det A(\xi))E$, from (2.21), (2.20) and (2.22) we see that for $x \neq y$

$$A(\partial_x)D(x, y) = A(\partial_x)A^*(\partial_x)t(x, y) = (\det A)(\partial_x)t(x, y) = 0.$$

Also, using (2.25) and expliciting the individual components, we easily convince ourselves that

$$A(\partial_x)P(x, y; n) = \bigl(T(\partial_y; n)\bigl(A(\partial_x)D(x, y)\bigr)^*\bigr)^* = 0.$$

2.9. Theorem (Somigliana representation formula). *If the (3×1)-matrix $u \in C^2(S^+) \cap C^1(\bar{S}^+)$ is a solution of the homogeneous system (2.8), then*

$$\phi(x)u(x) = \int_{\partial S} [D(x, y)T(\partial_y)u(y) - P(x, y)u(y)]\, ds(y), \tag{2.30}$$

where

$$\phi(x) = \begin{cases} 1, & x \in S^+, \\ \frac{1}{2}, & x \in \partial S, \\ 0, & x \in S^-. \end{cases} \tag{2.31}$$

Proof. Let $x \in S^+$, and let $\sigma_{x,\delta} \subset S^+$ be a disk with centre at x and radius δ sufficiently small. Applying Theorem 2.7 in $S^+ \setminus \sigma_{x,\delta}$, with $\tilde{u}$ replaced in turn by each column of D, and making use of Theorem 2.8, we find that

$$\int_{\partial S} \left[D(x,y)T(\partial_y)u(y) - P(x,y)u(y) \right] ds(y)$$
$$= \int_{\partial \sigma_{x,\delta}} \left[D(x,y)T(\partial_y)u(y) - P(x,y)u(y) \right] ds(y), \quad (2.32)$$

where $\partial \sigma_{x,\delta}$ is the boundary of $\sigma_{x,\delta}$.

By (2.27),

$$\int_{\partial \sigma_{x,\delta}} D(x,y)T(\partial_y)u(y)\, ds(y) = O(\delta \ln \delta).$$

From (2.28) we see that for $y \in \partial \sigma_{x,\delta}$

$$P(x,y) = O(\delta^{-1}),$$
$$\int_{\partial \sigma_{x,\delta}} P(x,y)\, ds(y) = -E + O(\delta \ln \delta).$$

Consequently, since $u(x) - u(y) = O(\delta)$,

$$\int_{\partial \sigma_{x,\delta}} P(x,y)u(y)\, ds(y)$$
$$= \int_{\partial \sigma_{x,\delta}} P(x,y)[u(y) - u(x)]\, ds(y) + \left[\int_{\partial \sigma_{x,\delta}} P(x,y)\, ds(y) \right] u(x)$$
$$= -u(x) + O(\delta).$$

The first part of the assertion now follows from (2.32) if we let $\delta \to 0$.

The case when $x \in \partial S$ is handled in a similar way, with $\sigma_{x,\delta}$ replaced by $\sigma_{x,\delta} \cap S^+$ and $\partial \sigma_{x,\delta}$ by its part lying in S^+. As was remarked in the proof of Theorem 1.46, for δ small the length of the latter is $\pi\delta + O(\delta^2)$, which yields the required formula.

The result when $x \in S^-$ is obtained directly from (2.32).

2.3. The case of the exterior domain

For y fixed and $|x| \to \infty$ we have

$$|x-y|^{-2} = |x|^{-2} + 2\langle x, y\rangle |x|^{-4} - |y|^2|x|^{-4} + 4\langle x, y\rangle^2 |x|^{-6} + O(|x|^{-5}),$$

$$\ln|x-y| = \ln|x| - \langle x, y\rangle |x|^{-2} + \tfrac{1}{2}|y|^2|x|^{-2} - \langle x, y\rangle^2 |x|^{-4} \qquad (2.33)$$
$$+ \langle x, y\rangle |y|^2|x|^{-4} - \tfrac{4}{3}\langle x, y\rangle^3 |x|^{-6} + O(|x|^{-4}),$$

and [1]

$$K_0(|x-y|/h) = O(|x|^{-1/2} e^{-|x|}). \qquad (2.34)$$

Then from (2.19), (2.21), (2.23), (2.25), (2.33), and (2.34) we obtain the asymptotic relations

$$D_{33} = O(|x|^2 \ln|x|),$$
$$D_{\alpha 3}, D_{3\alpha} = O(|x| \ln|x|),$$
$$D_{11}, D_{22} = O(\ln|x|),$$
$$D_{12}, D_{21} = O(1), \qquad (2.35)$$
$$P_{3\alpha} = O(\ln|x|),$$
$$P_{\alpha\beta}, P_{33} = O(|x|^{-1}),$$
$$P_{\alpha 3} = O(|x|^{-2}).$$

This means that we cannot obtain analogues of Theorems 2.5 and 2.9 in S^- without restrictions on the behaviour of u at infinity.

Let $\mathcal{A}$ be the set of (3×1)-matrices u admitting an asymptotic expansion of the form

$$u_1(r, \theta) = r^{-1}\left[m_0 \sin\theta + 2m_1 \cos\theta - m_0 \sin 3\theta + (m_2 - m_1)\cos 3\theta\right]$$
$$+ r^{-2}\left[(2m_3 + m_4)\sin 2\theta + m_5 \cos 2\theta - 2m_3 \sin 4\theta + 2m_6 \cos 4\theta\right]$$
$$+ r^{-3}\left[2m_7 \sin 3\theta + 2m_8 \cos 3\theta + 3(m_9 - m_7)\sin 5\theta\right.$$
$$\left. + 3(m_{10} - m_8)\cos 5\theta\right] + O(r^{-4}),$$

$$u_2(r, \theta) = r^{-1}\left[2m_2 \sin\theta + m_0 \cos\theta + (m_2 - m_1)\sin 3\theta + m_0 \cos 3\theta\right]$$
$$+ r^{-2}\left[(2m_6 + m_5)\sin 2\theta - m_4 \cos 2\theta + 2m_6 \sin 4\theta + 2m_3 \cos 4\theta\right]$$
$$+ r^{-3}\left[2m_{10} \sin 3\theta - 2m_9 \cos 3\theta + 3(m_{10} - m_8)\sin 5\theta\right. \quad (2.36)$$
$$\left. + 3(m_7 - m_9)\cos 5\theta\right] + O(r^{-4}),$$

$$u_3(r, \theta) = -(m_1 + m_2)\ln r - \left[m_1 + m_2 + m_0 \sin 2\theta + (m_1 - m_2)\cos 2\theta\right]$$
$$+ r^{-1}\left[(m_3 + m_4)\sin\theta + (m_5 + m_6)\cos\theta - m_3 \sin 3\theta + m_6 \cos 3\theta\right]$$
$$+ r^{-2}\left[m_{11} \sin 2\theta + m_{12} \cos 2\theta + (m_9 - m_7)\sin 4\theta\right.$$
$$\left. + (m_{10} - m_8)\cos 4\theta\right] + O(r^{-3}),$$

where (r, θ) are polar coordinates and $m_1, \ldots, m_{12}$ arbitrary constants. We also introduce the set

$$\mathcal{A}^* = \{u^* \mid u^* = u + u_0, \ u \in \mathcal{A}, \ u_0 \text{ is of the form } (2.16)\}.$$

2.10. Remark. In view of (2.12), $\mathcal{A}$ and $\mathcal{A}^*$ are classes of finite energy functions.

For simplicity, throughout what follows we consider only the homogeneous system (2.8), that is,

$$Au = 0. \quad (2.37)$$

This is done without loss of generality since, as in [24], if F is sufficiently smooth, then (2.8) can be reduced to (2.37) by means of a particular solution constructed in the form of a Newtonian potential.

2.11. Theorem (Somigliana representation formula). *If the (3×1)-matrix $u \in C^2(S^-) \cap C^1(\bar{S}^-) \cap \mathcal{A}$ is a solution of (2.37), then*

$$[1 - \phi(x)]u(x) = -\int_{\partial S} [D(x, y)T(\partial_y)u(y) - P(x, y)u(y)]\, ds(y),$$

where ϕ is defined by (2.31).

Proof. Consider a circle Γ_R with the centre at x and radius R sufficiently large so that ∂S lies inside Γ_R. With the origin of polar coordinates at x, from (2.11), (2.19), (2.21), (2.23), (2.25), and (2.33)–(2.36) we find that for $y = (R, \theta) \in \partial \Gamma_R$

$$T_{3i}u_i = R^{-3}\left[(m_7 + m_9 - 2m_{11})\sin 2\theta + (m_8 + m_{10} - 2m_{12})\cos 2\theta\right]$$
$$+ O(R^{-4}),$$

$$(D_{3\alpha}T_{\alpha i} - P_{3i})u_i = [4(\lambda + 2\mu)]^{-1} R^{-1}(4\lambda \ln R + 3\lambda + 2\mu)$$
$$\times \left[m_0 \sin 2\theta + 2(m_2 - m_1)\cos 2\theta\right] + O(R^{-2}\ln R),$$

$$(D_{\alpha i}T_{ij} - P_{\alpha j})u_j = O(R^{-2}\ln R).$$

Consequently,

$$\int_{\partial \Gamma_R} [D(x, y)Tu(y) - P(x, y)u(y)]\, ds(y) = O(R^{-1}\ln R),$$

and the result is obtained by applying Theorem 2.9 in $S^- \cap \Gamma_R$ and letting $R \to \infty$.

2.12. Theorem (Betti formula). *If $u \in C^2(S^-) \cap C^1(\bar{S}^-) \cap \mathcal{A}^*$ is a solution of (2.37), then*

$$2 \int_{S^-} E(u, u)\, da = -\int_{\partial S} u^T Tu\, ds.$$

Proof. The required formula is obtained via the procedure used in the proof of Theorem 2.11—this time in conjunction with Theorem 2.5—after noting that for R large

$$\begin{aligned} T_{\alpha i} u_i &= O(R^{-2}), \\ T_{3i} u_i &= O(R^{-3}), \end{aligned} \quad (2.38)$$

which means that $u^T Tu = O(R^{-2})$ for $u \in \mathcal{A}^*$.

2.4. Uniqueness of regular solutions

Let $\mathcal{P}(x)$, $\mathcal{Q}(x)$, $\mathcal{R}(x)$, and $\mathcal{S}(x)$ be continuous (3×1)-matrices prescribed on ∂S. We consider the following interior and exterior Dirichlet and Neumann boundary value problems:

(D$^+$) *Find* $u \in C^2(S^+) \cap C^1(\bar{S}^+)$ *satisfying*

$$\begin{aligned} Au(x) &= 0, & x &\in S^+, \\ u(x) &= \mathcal{P}(x), & x &\in \partial S. \end{aligned} \quad (2.39)$$

(N$^+$) *Find* $u \in C^2(S^+) \cap C^1(\bar{S}^+)$ *satisfying*

$$\begin{aligned} Au(x) &= 0, & x &\in S^+, \\ Tu(x) &= \mathcal{Q}(x), & x &\in \partial S. \end{aligned} \quad (2.40)$$

(D$^-$) *Find* $u \in C^2(S^-) \cap C^1(\bar{S}^-) \cap \mathcal{A}^*$ *satisfying*

$$\begin{aligned} Au(x) &= 0, & x &\in S^-, \\ u(x) &= \mathcal{R}(x), & x &\in \partial S. \end{aligned} \quad (2.41)$$

(N^-) *Find* $u \in C^2(S^-) \cap C^1(\bar{S}^-) \cap \mathcal{A}$ *satisfying*

$$Au(x) = 0, \quad x \in S^-,$$
$$Tu(x) = \mathcal{S}(x), \quad x \in \partial S. \tag{2.42}$$

2.13. Definition. A solution as stated above is called a *regular solution* of the corresponding problem.

2.14. Remark. The condition that $u \in C^1(\bar{S}^+)$ or $u \in C^1(\bar{S}^-)$ is necessary even in the case of the Dirichlet problems, to ensure the applicability of the Betti formula.

2.15. Theorem. *(i)* (D^+), (D^-) *and* (N^-) *have at most one regular solution.*

(ii) Any two regular solutions of (N^+) *differ by a* (3×1)-*matrix of the form* (2.16).

Proof. (i) The difference u of two regular solutions of (D^+) satisfies (2.39) with $\mathcal{P} = 0$, therefore, by Theorem 2.6 and the fact that $E(u,u)$ is a positive quadratic form,

$$E(u, u) = 0 \quad \text{in} \quad S^+.$$

From Theorem 2.3 it now follows that u is of the form (2.16) in $\bar{S}^+$. Since $u = 0$ on ∂S, we deduce that $u(x) = 0$, $x \in \bar{S}^+$.

The same argument, but based on Theorem 2.12 instead of Theorem 2.6, is used to prove the result for (D^-).

If u is the difference of two regular solutions of (N^-), then, as above, we conclude that u is of the form (2.16) in S^-. However, since $u \in \mathcal{A}$, we see from (2.36) that $u = 0$.

(ii) As in the case of (D^+), we find that the difference of two regular solutions of (N^+) is of the form (2.16).

2.5. Elastic potentials with smooth densities

Let $p = (p_{ij})$, $i = 1, \ldots, n$, $j = 1, \ldots, m$, be a matrix, X a space of scalar functions, and L a scalar operator on X. In what follows we write $p \in X$ if $p_{ij} \in X$, $i = 1, \ldots, n$, $j = 1, \ldots, m$, and $Lp = (Lp_{ij})$.

We introduce the elastic single layer potential

$$(V(\varphi))(x) = \int_{\partial S} D(x, y)\varphi(y)\, ds(y) \qquad (2.43)$$

and the elastic double layer potential

$$(W(\varphi))(x) = \int_{\partial S} P(x, y)\varphi(y)\, ds(y), \qquad (2.44)$$

where φ is a density (3×1)-matrix.

As in the case of harmonic potentials, when specifying the density is not essential we use the simpler notation $V(x)$ and $W(x)$.

2.16. Theorem. *If $\varphi \in C(\partial S)$, then $V(\varphi)$ and $W(\varphi)$ are analytic and satisfy (2.37) in $S^+ \cup S^-$.*

Proof. Clearly, $V(\varphi)$ and $W(\varphi)$ are twice continuously differentiable at any $x \notin \partial S$ and, by Theorem 2.8, are solutions of (2.37). Their analyticity follows in the usual way (see, for example, [26]).

2.17. Theorem. *If $\varphi \in C(\partial S)$, then*

(i) $W(\varphi) \in \mathcal{A}$;

(ii) $V(\varphi) \in \mathcal{A}$ *if and only if*

$$p_\alpha = \int_{\partial S} (\varphi_\alpha - x_\alpha \varphi_3)\, ds = 0,$$

$$p = \int_{\partial S} \varphi_3\, ds = 0.$$

Proof. (i) The expansion (2.36) for $W(\varphi)$ is ontained from (2.44), (2.25), (2.11), (2.19), (2.21), (2.23), (2.33), and (2.34).

(ii) Using the same formulae as above, from (2.43) we find that, as $r = |x| \to \infty$,

$$V_1(r, \theta) = -a_2\mu^2 \big[pr(2\ln r + 1)\cos\theta + p_1(2\ln r + 2 + \cos 2\theta) + p_2 \sin 2\theta\big]$$
$$+ \tilde{V}_1(r, \theta),$$

$$V_2(r, \theta) = -a_2\mu^2 \big[pr(2\ln r + 1)\sin\theta + p_2(2\ln r + 2 - \cos 2\theta) + p_1 \sin 2\theta\big]$$
$$+ \tilde{V}_2(r, \theta),$$

$$V_3(r, \theta) = a_2\mu p\big[\mu r^2 \ln r - 4h^2(\lambda + 2\mu)\ln r - 4h^2(\lambda + 3\mu)\big]$$
$$+ a_2\mu(p_1 \cos\theta + p_2 \sin\theta)\big[\mu r(2\ln r + 1) - 4h^2(\lambda + 2\mu)r^{-1}\big]$$
$$+ \tilde{V}_3(r, \theta),$$

where $\tilde{V} \in \mathcal{A}$, and the result follows immediately.

2.18. Remark. The requirement that the solutions of the exterior boundary value problems should belong to $\mathcal{A}$ or $\mathcal{A}^*$ is justified by the fact that such solutions will be sought in the form of single or double layer potentials.

In view of Theorem 2.16, to investigate the continuity and differentiability of V and W in $\bar{S}^+$ and $\bar{S}^-$ it suffices to consider their behaviour in $\bar{S}_0^+$ and $\bar{S}_0^-$.

2.19. Theorem. *If $\varphi \in C(\partial S)$, then $V(\varphi) \in C^{0,\alpha}(\mathbf{R}^2)$ for any index $\alpha \in (0, 1)$.*

Proof. From (2.43) and (2.27) we see that for $x \in S_0 \setminus \partial S$

$$V(\varphi) = \frac{1}{2\pi}\left(a_4 E_{\gamma\gamma} + \frac{1}{\mu}E_{33}\right)v(\varphi) - 2a_2\mu^2 E_{\alpha\beta}v^b_{\alpha\beta}(\varphi) + \tilde{V}(\varphi),$$

where v and $v_{\alpha\beta}^b$ are defined by (1.35) and (1.49), respectively, and

$$\tilde{V}(x) = \int_{\partial S} \tilde{D}(x,y)\varphi(y)\,ds(y).$$

The result now follows from Theorems 1.45, 1.55 and 1.53.

2.20. Theorem. *If $\varphi \in C^{0,\alpha}(\partial S)$, $\alpha \in (0,1]$, then $W(\varphi)$ has $C^{0,\beta}$-extensions W^+ and W^- to $\bar{S}^+$ and $\bar{S}^-$, respectively, with $\beta = \alpha$ for $\alpha \in (0,1)$ and any $\beta \in (0,1)$ for $\alpha = 1$. These extensions are given by*

$$W^+(x) = \begin{cases} W(x), & x \in S_0^+, \\ -\tfrac{1}{2}\varphi(x) + W_0(x), & x \in \partial S, \end{cases}$$

$$W^-(x) = \begin{cases} W(x), & x \in S_0^-, \\ \tfrac{1}{2}\varphi(x) + W_0(x), & x \in \partial S, \end{cases} \quad (2.45)$$

where

$$W_0(x) = \int_{\partial S} P(x,y)\varphi(y)\,ds(y), \quad x \in \partial S,$$

the integral being understood as principal value.

Proof. As in the proof of Theorem 2.19, from (2.44) and (2.28) we find that for $x \in S_0 \setminus \partial S$

$$W(\varphi) = -\frac{1}{2\pi}\Big[\mu'\varepsilon_{\alpha\beta}E_{\alpha\beta}v^f(\varphi) - w(\varphi) - (\lambda' + \mu')\varepsilon_{\alpha\gamma}E_{\gamma\beta}v_{\alpha\beta}^e(\varphi)$$

$$+ \frac{1}{2}\varepsilon_{\alpha\beta}\Big(\lambda'E_{3\beta} + \frac{1}{h^2}E_{\beta 3}\Big)v_\alpha^c(\varphi) - \frac{1}{2}E_{3\alpha}v_\alpha^d(\varphi)\Big]$$

$$+ \tilde{W}(\varphi), \quad (2.46)$$

where v_α^c, v_α^d, $v_{\alpha\beta}^e$, v^f, and w are defined by (1.50), (1.51), (1.52), (1.58), and (1.36), respectively,

$$\tilde{W}(x) = \int_{\partial S} \tilde{P}(x, y)\varphi(y)\, ds(y),$$

and the kernel $\tilde{P}(x, y)$ satisfies the conditions of Theorem 1.53 with any $\gamma \in (0, 1)$.

The assertion now follows from (2.46), Theorems 1.55, 1.56, 1.58, 1.53, and 1.46, and Remarks 1.59 and 1.47.

2.21. Theorem. *If $\varphi \in C^{0,\alpha}(\partial S)$, $\alpha \in (0, 1]$, then the first order derivatives of $V(\varphi)$ in S^+ and S^- have $C^{0,\beta}$-extensions to $\bar{S}^+$ and $\bar{S}^-$, respectively, with $\beta = \alpha$ for $\alpha \in (0, 1)$ and any $\beta \in (0, 1)$ for $\alpha = 1$. These extensions are given by*

$$(V_{,\gamma})^+(x) = \begin{cases} V_{,\gamma}(x), & x \in S_0^+, \\ \frac{1}{2} f^\gamma(x) + (V_{,\gamma})_0(x), & x \in \partial S, \end{cases}$$

$$(V_{,\gamma})^-(x) = \begin{cases} V_{,\gamma}(x), & x \in S_0^-, \\ -\frac{1}{2} f^\gamma(x) + (V_{,\gamma})_0(x), & x \in \partial S, \end{cases}$$

where

$$\begin{aligned} f_\alpha^\gamma &= 4a_2\mu\big[(\lambda + 2\mu)\delta_{\alpha\beta} - (\lambda + \mu)\nu_\alpha\nu_\beta\big]\nu_\gamma\varphi_\beta, \\ f_3^\gamma &= \frac{1}{2\pi\mu}\nu_\gamma\varphi_3, \end{aligned} \quad (2.47)$$

and

$$(V_{,\gamma})_0(x) = \int_{\partial S} \frac{\partial}{\partial x_\gamma} D(x, y)\varphi(y)\, ds(y), \quad x \in \partial S,$$

the integral being understood as principal value.

Proof. For $x \in S_0 \setminus \partial S$ and $y \in \partial S$ we write

$$\begin{aligned} D_{\alpha\beta} &= D^*_{\alpha\beta} + \tilde{D}_{\alpha\beta}, \\ D_{33} &= D^*_{33} + \tilde{D}_{33}, \end{aligned} \quad (2.48)$$

where, by (2.28),

$$D^*_{\alpha\beta}(x, y) = 2a_2\mu\Big[-(\lambda + 3\mu)\delta_{\alpha\beta}\ln|x - y|$$
$$+ (\lambda + \mu)\frac{(x_\alpha - y_\alpha)(x_\beta - y_\beta)}{|x - y|^2}\Big], \quad (2.49)$$

$$D^*_{33}(x, y) = -\frac{1}{2\pi\mu}\ln|x - y|,$$

and $\tilde{D}_{\alpha\beta}$ and $\tilde{D}_{33}$ satisfy the conditions of Theorem 1.53 with any $\gamma \in (0, 1)$. From (2.21), (2.19) and (2.23) it follows that

$$D_{\alpha3} = -D_{3\alpha} = -a_2\mu^2(x_\alpha - y_\alpha)(2\ln|x - y| + 1), \quad (2.50)$$

which also satisfies the conditions of Theorem 1.53 with any $\gamma \in (0, 1)$.

Using (1.7) and the equality

$$\varepsilon_{\beta\sigma}\varepsilon_{\mu\rho} = \delta_{\beta\mu}\delta_{\sigma\rho} - \delta_{\beta\rho}\delta_{\sigma\mu}, \quad (2.51)$$

we find that

$$\varepsilon_{\beta\sigma}\frac{\partial}{\partial s(y)}\frac{(x_\alpha - y_\alpha)(x_\sigma - y_\sigma)}{|x - y|^2}$$
$$= \Big[\nu_\beta(y)\frac{\partial}{\partial y_\sigma} - \nu_\sigma(y)\frac{\partial}{\partial y_\beta}\Big]\frac{(x_\alpha - y_\alpha)(x_\sigma - y_\sigma)}{|x - y|^2}$$
$$= -\delta_{\alpha\beta}\frac{\partial}{\partial\nu(y)}\ln|x - y| - 2\nu_\sigma(y)\frac{(x_\alpha - y_\alpha)(x_\beta - y_\beta)(x_\sigma - y_\sigma)}{|x - y|^4}. \quad (2.52)$$

Next, (2.52) and (1.21) yield

$$\frac{\partial}{\partial\nu(y)}\frac{(x_\alpha - y_\alpha)(x_\beta - y_\beta)}{|x - y|^2}$$
$$= -\nu_\alpha(y)\frac{x_\beta - y_\beta}{|x - y|^2} - \nu_\beta(y)\frac{x_\alpha - y_\alpha}{|x - y|^2}$$
$$+ 2\nu_\sigma(y)\frac{(x_\alpha - y_\alpha)(x_\beta - y_\beta)(x_\sigma - y_\sigma)}{|x - y|^4}$$

$$= \left[\nu_\alpha(y)\frac{\partial}{\partial y_\beta} + \nu_\beta(y)\frac{\partial}{\partial y_\alpha}\right]\ln|x-y| - \varepsilon_{\beta\sigma}\frac{\partial}{\partial s(y)}\frac{(x_\alpha - y_\alpha)(x_\sigma - y_\sigma)}{|x-y|^2}$$

$$- \delta_{\alpha\beta}\frac{\partial}{\partial \nu(y)}\ln|x-y|$$

$$= \left[\nu_\alpha(y)\tau_\beta(y) + \nu_\beta(y)\tau_\alpha(y)\right]\frac{\partial}{\partial s(y)}\ln|x-y|$$

$$+ \left[2\nu_\alpha(y)\nu_\beta(y) - \delta_{\alpha\beta}\right]\frac{\partial}{\partial \nu(y)}\ln|x-y|$$

$$- \varepsilon_{\beta\sigma}\frac{\partial}{\partial s(y)}\frac{(x_\alpha - y_\alpha)(x_\sigma - y_\sigma)}{|x-y|^2}. \qquad (2.53)$$

Finally, from (2.53), (2.49), (1.7), and (1.21) we obtain

$$\frac{\partial}{\partial x_\gamma}D^*_{\alpha\beta}(x,y) = -\frac{\partial}{\partial y_\gamma}D^*_{\alpha\beta}(y,x)$$

$$= 2a_2\mu\bigg\{(\lambda + 3\mu)\delta_{\alpha\beta}\left[\varepsilon_{\sigma\gamma}\nu_\sigma(y)\frac{\partial}{\partial s(y)} + \nu_\gamma(y)\frac{\partial}{\partial \nu(y)}\right]\ln|x-y|$$

$$+ (\lambda + \mu)\left[\varepsilon_{\gamma\sigma}\nu_\sigma(y)\frac{\partial}{\partial s(y)} - \nu_\gamma(y)\frac{\partial}{\partial \nu(y)}\right]\frac{(x_\alpha - y_\alpha)(x_\beta - y_\beta)}{|x-y|^2}\bigg\}$$

$$= 2a_2\mu\bigg\{2\nu_\gamma(y)\left[(\lambda + 2\mu)\delta_{\alpha\beta} - (\lambda + \mu)\nu_\alpha(y)\nu_\beta(y)\right]\frac{\partial}{\partial \nu(y)}\ln|x-y|$$

$$+ \left[(\lambda + 3\mu)\delta_{\alpha\beta}\varepsilon_{\sigma\gamma}\nu_\sigma(y)\right.$$

$$- (\lambda + \mu)\left(\varepsilon_{\sigma\beta}\nu_\alpha(y) + \varepsilon_{\sigma\alpha}\nu_\beta(y)\right)\nu_\sigma(y)\nu_\gamma(y)\bigg]\frac{\partial}{\partial s(y)}\ln|x-y|$$

$$- (\lambda + \mu)\left[\varepsilon_{\sigma\gamma}\nu_\sigma(y)\frac{\partial}{\partial s(y)}\frac{(x_\alpha - y_\alpha)(x_\beta - y_\beta)}{|x-y|^2}\right.$$

$$\left.+ \varepsilon_{\sigma\beta}\nu_\gamma(y)\frac{\partial}{\partial s(y)}\frac{(x_\alpha - y_\alpha)(x_\beta - y_\beta)}{|x-y|^2}\right]\bigg\}.$$

Similarly,

$$\frac{\partial}{\partial x_\gamma}D^*_{33}(x,y) = \frac{1}{2\pi\mu}\left[\nu_\gamma(y)\frac{\partial}{\partial \nu(y)} + \varepsilon_{\sigma\gamma}\nu_\sigma(y)\frac{\partial}{\partial s(y)}\right]\ln|x-y|.$$

In view of (2.43) and the above calculation, for $x \in S_0 \setminus \partial S$ and $y \in \partial S$ we can now write

$$V_{,\gamma}(\varphi) = -\pi^{-1}w(f^\gamma) + v(g^\gamma) + v^b_{\alpha\gamma}(p^\alpha)$$
$$+ v^e_{\alpha\beta}(q^{\alpha\beta\gamma}) + v^f(r^\gamma) + \mathcal{V}^\gamma. \qquad (2.54)$$

Here w, v, $v^b_{\alpha\gamma}$, $v^e_{\alpha\beta}$, and v^f are defined by (1.36), (1.35), (1.49), (1.52), and (1.58), respectively,

$$\mathcal{V}^\gamma(x) = \int_{\partial S} \mathcal{F}(x, y) t^\gamma(y) \, ds(y),$$

the densities f^γ (given by (2.47)), g^γ, p^α, $q^{\alpha\beta\gamma}$, r^γ, and t^γ are (3×1)-matrices of class $C^{0,\alpha}(\partial S)$, and $\mathcal{F}(x, y)$ is a proper δ-singular kernel in S_0 for any $\delta \in (0, 1)$. The assertion now follows from Theorems 1.46, 1.45, 1.55, 1.56, 1.58, and 1.30.

2.22. Remark. From Theorems 2.19 and 2.21 we conclude that if $\varphi \in C^{0,\alpha}(\partial S)$, $\alpha \in (0, 1]$, then the restrictions of $V(\varphi)$ to $\bar{S}^+$ and $\bar{S}^-$ belong respectively to $C^{1,\beta}(\bar{S}^+)$ and $C^{1,\beta}(\bar{S}^-)$, with $\beta = \alpha$ for $\alpha \in (0, 1)$ and any $\beta \in (0, 1)$ for $\alpha = 1$. Denoting these two functions by V^+ and V^-, we can write

$$V^+_{,\gamma}(x) = (V_{,\gamma})^+(x), \quad x \in \bar{S}^+,$$
$$V^-_{,\gamma}(x) = (V_{,\gamma})^-(x), \quad x \in \bar{S}^-.$$

2.23. Corollary. *If $\varphi \in C^{0,\alpha}(\partial S)$, $\alpha \in (0, 1]$, then the restrictions of $TV(\varphi)$ to S^+ and S^- have $C^{0,\beta}$-extensions to $\bar{S}^+$ and $\bar{S}^-$, respectively, with $\beta = \alpha$ for $\alpha \in (0, 1)$ and any $\beta \in (0, 1)$ for $\alpha = 1$. These extensions are given by*

$$(TV)^+(x) = \begin{cases} TV(x), & x \in S_0^+, \\ \tfrac{1}{2}\varphi(x) + (TV)_0(x), & x \in \partial S, \end{cases}$$
$$(TV)^-(x) = \begin{cases} TV(x), & x \in S_0^-, \\ -\tfrac{1}{2}\varphi(x) + (TV)_0(x), & x \in \partial S, \end{cases} \quad (2.55)$$

where
$$(TV)_0(x) = \int_{\partial S} T(\partial_x) D(x,y) \varphi(y)\, ds(y),$$

the integral being understood as principal value.

Proof. By (2.11), for $x \in S_0 \setminus \partial S$ we have
$$\begin{aligned}(TV)_\alpha &= h^2\big[\lambda \nu_\alpha V_{\beta,\beta} + \mu \nu_\beta (V_{\beta,\alpha} + V_{\alpha,\beta})\big], \\ (TV)_3 &= \mu \nu_\alpha (V_\alpha + V_{3,\alpha}),\end{aligned} \quad (2.56)$$

and the $C^{0,\beta}$-extendability of TV to $\bar S_0^+$ and $\bar S_0^-$ follows from Theorem 2.21. For $x \in \partial S$ the same theorem and (2.56) yield

$$(TV)_\alpha^\pm = \pm h^2\big[\lambda \nu_a f_\beta^\beta + \mu \nu_\beta (f_\alpha^\beta + f_\beta^\alpha)\big] + (TV)_0 = \pm \tfrac{1}{2}\varphi_\alpha + ((TV)_0)_\alpha,$$
$$(TV)_3^\pm = \mu \nu_\alpha f_3^\alpha + ((TV)_0)_3 = \pm \tfrac{1}{2}\varphi_3 + ((TV)_0)_3,$$

which completes the proof.

2.24. Remark. In view of Remark 2.22, we can write
$$\begin{aligned}TV^+(x) &= (TV)^+(x), \quad x \in \bar S^+, \\ TV^-(x) &= (TV)^-(x), \quad x \in \bar S^-.\end{aligned} \quad (2.57)$$

2.25. Theorem. *If $\varphi \in C^{1,\alpha}(\partial S)$, $\alpha \in (0,1]$, then the restrictions of $W(\varphi)$ to S^+ and S^- have $C^{1,\beta}$-extensions W^+ and W^- to $\bar S_0^+$ and $\bar S_0^-$, respectively, with $\beta = \alpha$ for $\alpha \in (0,1)$ and any $\beta \in (0,1)$ for $\alpha = 1$. These extensions are given by (2.45) and satisfy $TW^+ = TW^-$ on ∂S.*

Proof. Let $u(x)$ be a continuously differentiable (3×1)-matrix. From (2.11), (1.7) and the easily verified equality

$$\nu_\beta u_{\beta,\gamma} - \nu_\gamma u_{\beta,\beta} = \varepsilon_{\beta\gamma} \frac{\partial}{\partial s} u_\beta$$

we obtain

$$T_{\gamma\beta} u_\beta = h^2 \left[\lambda \nu_\gamma u_{\beta,\beta} + \mu \nu_\beta (u_{\beta,\gamma} + u_{\gamma,\beta}) \right]$$

$$= h^2 \left\{ 2\mu \varepsilon_{\beta\gamma} \frac{\partial}{\partial s} u_\beta \right.$$

$$\left. + \left[(\lambda + \mu)\nu_\gamma \frac{\partial}{\partial x_\beta} - \mu \varepsilon_{\beta\gamma} \frac{\partial}{\partial s} + \mu \delta_{\beta\gamma} \frac{\partial}{\partial \nu} \right] u_\beta \right\}. \quad (2.58)$$

Using (2.49), (2.50) and (2.51), after a lengthy but straightforward calculation we find that for $x \in S_0 \setminus \partial S$ and $y \in \partial S$

$$\left[(\lambda + \mu)\nu_\gamma(y) \frac{\partial}{\partial y_\beta} - \mu \varepsilon_{\beta\gamma} \frac{\partial}{\partial s(y)} + \mu \delta_{\beta\gamma} \frac{\partial}{\partial \nu(y)} \right] D^*_{\alpha\beta}(x, y)$$

$$= \frac{1}{2\pi} \left[-\delta_{\alpha\gamma} \frac{\partial}{\partial \nu(y)} \ln |x - y| + \varepsilon_{\alpha\gamma} \frac{\partial}{\partial s(y)} \ln |x - y| \right].$$

Consequently, using (2.58), (2.24) and integration by parts, we see that

$$\int_{\partial S} [T_{\gamma\beta}(\partial_y) D^*_{\beta\alpha}(y, x)] \varphi_\gamma(y) \, ds(y)$$

$$= 2h^2 \mu \varepsilon_{\gamma\beta} \int_{\partial S} D^*_{\alpha\beta}(x, y) \varphi'_\gamma(y) \, ds(y)$$

$$- \frac{1}{2\pi} \int_{\partial S} \left[\frac{\partial}{\partial \nu(y)} \ln |x - y| \right] \varphi_\alpha(y) \, ds(y)$$

$$- \frac{1}{2\pi} \varepsilon_{\alpha\gamma} \int_{\partial S} (\ln |x - y|) \varphi'_\gamma(y) \, ds(y).$$

Similarly,

$$\int_{\partial S} [T_{3\beta}(\partial_y) D_{\beta\alpha}(y, x)] \varphi_3(y) \, ds(y) = \mu \int_{\partial S} D_{\alpha\beta}(x, y) \nu_\beta(y) \varphi_3(y) \, ds(y)$$

and, by (2.50),

$$\int_{\partial S} [T_{33}(\partial_y) D_{3\alpha}(y, x)] \varphi_3(y) \, ds(y)$$

$$= 2a_2 \mu^3 \bigg\{ \int_{\partial S} (\ln|x-y|) \nu_\alpha(y) \varphi_3(y) \, ds(y)$$

$$- \int_{\partial S} \bigg[\frac{\partial}{\partial \nu(y)} \ln|x-y| \bigg] (x_\alpha - y_\alpha) \varphi_3(y) \, ds(y) \bigg\} + c_\alpha,$$

where c_α are combinations of λ and μ.

By means of the same procedure and (2.24), we arrive at

$$\int_{\partial S} [T_{\gamma\beta}(\partial_y) D_{\beta 3}(y, x)] \varphi_\gamma(y) \, ds(y)$$

$$= 2h^2 \mu \bigg\{ \varepsilon_{\gamma\beta} \int_{\partial S} D_{3\beta}(x, y) \varphi'_\gamma(y) \, ds(y)$$

$$- a_2 \mu(2\lambda + 3\mu) \int_{\partial S} (\ln|x-y|) \nu_\gamma(y) \varphi_\gamma(y) \, ds(y)$$

$$+ a_2 \mu^2 \int_{\partial S} \bigg[\frac{\partial}{\partial \nu(y)} \ln|x-y| \bigg] (x_\gamma - y_\gamma) \varphi_\gamma(y) \, ds(y)$$

$$- a_2 \mu^2 \varepsilon_{\gamma\beta} \int_{\partial S} (\ln|x-y|)(x_\beta - y_\beta) \varphi'_\gamma(y) \, ds(y) \bigg\} + c'_\gamma x_\gamma + c',$$

$$\int_{\partial S} [T_{3\beta}(\partial_y) D_{\beta 3}(y, x)] \varphi_3(y) \, ds(y) = \mu \int_{\partial S} D_{3\beta}(x, y) \nu_\beta(y) \varphi_3(y) \, ds(y),$$

$$\int_{\partial S} [T_{33}(\partial_y)D^*_{33}(y,x)]\varphi_3(y)\,ds(y) = -\frac{1}{2\pi}\int_{\partial S}\left[\frac{\partial}{\partial\nu(y)}\ln|x-y|\right]\varphi_3(y)\,ds(y),$$

where c'_γ and c' are combinations of λ and μ.

These relations, (2.43), (2.44), (2.25), and (2.48) yield

$$W(\varphi) = v\big(2h^2 a_2\mu^2\vartheta - (2\pi)^{-1}\sigma' - 2a_2\mu^3\varpi\big) + w\big((2\pi)^{-1}\varphi - 2a_2\mu^3\varrho\big) \\ + V(2h^2\mu\sigma' + \mu\varpi) + 2a_2\mu^3\tilde{W} + \mathcal{W}, \qquad (2.58)$$

where, as functions of x, the specified densities are

$$\vartheta = \big(0,\,0,\,(2\lambda+3\mu)\nu_\gamma\varphi_\gamma + \mu\varepsilon_{\gamma\beta}x_\gamma\varphi'_\beta\big)^{\mathrm{T}}, \\ \sigma = (-\varphi_2,\,\varphi_1,\,0)^{\mathrm{T}}, \\ \varpi = (\nu_1\varphi_3,\,\nu_2\varphi_3,\,0)^{\mathrm{T}}, \qquad (2.59) \\ \varrho = (x_1\varphi_3,\,x_2\varphi_3,\,-h^2 x_\gamma\varphi_\gamma)^{\mathrm{T}}, \\ \tilde{W} = \big(x_1 w(\varphi_3),\,x_2 w(\varphi_3),\,h^2 x_\gamma\big(v(\varepsilon_{\beta\gamma}\varphi'_\beta) - w(\varphi_\gamma)\big)\big)^{\mathrm{T}},$$

and $\mathcal{W}$ is a potential-type function whose kernel satisfies the conditions of Theorem 1.53 with any $\gamma \in (0,1)$. Since $\varphi \in C^{1,\alpha}(\partial S)$ and $\nu \in C^1(\partial S)$, we see immediately that

$$\vartheta \in C^{0,\alpha}(\partial S), \quad \varpi \in C^1(\partial S), \quad \sigma, \varrho \in C^{1,\alpha}(\partial S).$$

By Theorem 2.20, the restrictions of $\mathcal{W}$ to S^+ and S^- have $C^{0,\beta}$-extensions $\mathcal{W}^+$ and $\mathcal{W}^-$ to $\bar{S}^+$ and $\bar{S}^-$, respectively. On the other hand, by Theorems 1.48, 1.51, 2.21, and 1.53, the restrictions of $W_{,\gamma}$ to S^+ and S^- have $C^{0,\beta}$-extensions $(W_{,\gamma})^+$ and $(W_{,\gamma})^-$, given by (2.45), to $\bar{S}^+$ and $\bar{S}^-$. Since $W^+_{,\gamma}(x) = (W_{,\gamma})^+(x)$, $x \in S^+$, and $W^-_{,\gamma}(x) = (W_{,\gamma})^-(x)$, $x \in S^-$, the first part of the assertion follows from Theorem 1.17.

For the second part, first we deduce from (1.45) and (1.44) that

$$\left(\frac{\partial}{\partial x_\gamma}(w(\varphi))(x)\right)^\pm = \varepsilon_{\gamma\beta}\left\{\pm\pi\nu_\beta(x)\varphi'(x) + \int_{\partial S}\left(\frac{\partial}{\partial x_\beta}\ln|x-y|\right)\varphi'(y)\,ds(y)\right\}, \quad (2.60)$$

where the integral is understood as principal value. Next, we convince ourselves by direct verification that for $x \in \partial S$

$$\varepsilon_{\gamma\beta}\nu_\gamma(x_\beta\varphi_3)' = \varphi_3 + \varepsilon_{\gamma\beta}x_\beta\nu_\gamma\varphi_3',$$
$$\varepsilon_{\gamma\alpha}\nu_\gamma(x_\beta\varphi_3)' = -\nu_\alpha\nu_\beta\varphi_3 + \varphi_3\delta_{\alpha\beta} + \varepsilon_{\gamma\alpha}x_\beta\nu_\gamma\varphi_3',$$
$$\varepsilon_{\gamma\beta}\nu_\gamma(x_\alpha\varphi_3)' = -\nu_\alpha\nu_\beta\varphi_3 + \varphi_3\delta_{\alpha\beta} + \varepsilon_{\gamma\beta}x_\alpha\nu_\gamma\varphi_3',$$
$$\varepsilon_{\beta\alpha}\nu_\beta(x_\gamma\varphi_\gamma)' = \varphi_\alpha - \nu_\alpha\nu_\gamma\varphi_\gamma + \varepsilon_{\beta\alpha}x_\gamma\nu_\beta\varphi_\gamma',$$
$$\varepsilon_{\gamma\alpha}\nu_\beta\nu_\gamma\varphi_\beta' - \varepsilon_{\gamma\beta}\nu_\alpha\nu_\beta\varphi_\gamma' = \varepsilon_{\gamma\alpha}\varphi_\gamma'.$$

Finally, rewriting (2.58) in the form

$$W(\varphi) = V(2h^2\mu\sigma' + \mu\varpi) + \hat{W}, \quad (2.61)$$

starting from (2.11) and making use of (2.58), (1.43), (1.44), (2.60), (2.59), and the above relations, after another simple but rather lengthy computation we get

$$(T\hat{W}^\pm)_\alpha = (T\hat{W})_\alpha^\pm = (T_{\alpha\beta}\hat{W}_\beta)^\pm$$
$$= \mp\tfrac{1}{2}\mu[\nu_\alpha\varphi_3 + h^2(\varepsilon_{\gamma\beta}\nu_\alpha\nu_\beta\varphi_\gamma' + \varepsilon_{\gamma\alpha}\nu_\gamma\nu_\beta\varphi_\beta' + \varepsilon_{\gamma\alpha}\varphi_\gamma')] + \hat{W}_\alpha$$
$$= \mp\mu(\tfrac{1}{2}\nu_\alpha\varphi_3 + h^2\varepsilon_{\gamma\alpha}\varphi_\gamma') + \hat{W}_\alpha,$$
$$(T\hat{W}^\pm)_3 = (T\hat{W})_3^\pm = \hat{W}_3.$$

On the other hand, by Remark 2.24, Corollary 2.23, (2.61), and (2.59),

$$(TV^{\pm}(2h^2\mu\sigma' + \mu\varpi))_\alpha = (TV(2h^2\mu\sigma' + \mu\varpi))_\alpha^{\pm}$$
$$= \pm\mu(\tfrac{1}{2}\nu_\alpha\varphi_3 + h^2\varepsilon_{\gamma\alpha}\varphi'_\gamma) + ((TV)_0)_\alpha,$$
$$(TV^{\pm}(2h^2\mu\sigma' + \mu\varpi))_3 = (TV(2h^2\mu\sigma' + \mu\varpi))_\alpha^{\pm} = ((TV)_0)_3.$$

Consequently,

$$TW^+(x) = TW^-(x) = \hat{W}(x) + (TV)_0(x), \quad x \in \partial S,$$

which completes the proof of the theorem.

2.6. Elastic potentials with integrable densities

We denote by $L^p(\partial S)$, $p \geq 1$, the space of functions f on ∂S that are measurable and such that $|f|^p$ is Lebesgue integrable over ∂S. As is well known (see, for example, [41]), the space $L^p(\partial S)$ is complete with respect to the norm

$$\|f\|_p = \left\{ \int_{\partial S} |f(y)|^p \, ds(y) \right\}^{1/p}.$$

Also, every function $f \in L^1(\partial S)$ can be written in the form

$$f = f_1 - f_2, \tag{2.62}$$

where f_1 and f_2 are the limits almost everywhere of increasing sequences of step functions $\{\varphi_n^{(1)}\}_{n=1}^\infty$ and $\{\varphi_n^{(2)}\}_{n=1}^\infty$, respectively, for which the corresponding sequences of integrals $\{\int_{\partial S} \varphi_n\}_{n=1}^\infty$ and $\{\int_{\partial S} \psi_n\}_{n=1}^\infty$ are bounded.

2.26. Definition. Let $f \in L^1(\partial S)$. A point $x \in \partial S$ with the property that $\{\varphi_n^{(\alpha)}(x)\}_{n=1}^\infty \to f_\alpha(x)$, where the f_α are as in (2.62), is called a *Lebesgue point* for f.

2.27. Lemma. *If $f \in L^1(\partial S)$ and x is a Lebesgue point for f, then*

$$\lim_{\varepsilon \to 0} \frac{1}{\varepsilon} \int_s^{s+\varepsilon} f(t)\, dt = f(s).$$

The proof of this assertion is based on the decomposition (2.62) and Definition 2.26.

2.28. Theorem. *Let $k(x, y)$ be a proper γ-singular kernel in S_0, $\gamma \in [0, 1]$, let $x = \xi \mp \delta\nu(\xi) \in S_0^\pm$, $\xi \in \partial S$, $\delta > 0$, and suppose that*

$$\lim_{\delta \to 0} \int_{\partial S} k(x, y)\, ds(y) = l^\pm(\xi) \qquad (2.63)$$

and

$$\lim_{\delta \to 0} \int_{\partial S \setminus \Sigma_{\xi,\delta}} k(\xi, y)\, ds(y) = l(\xi), \qquad (2.64)$$

where $\Sigma_{\xi,\delta}$ is defined by (1.12). If $\varphi \in L^1(\partial S)$, then

$$\lim_{\delta \to 0} \left[\int_{\partial S} k(x, y)\varphi(y)\, ds(y) - \int_{\partial S \setminus \Gamma_{\xi,8\delta}} k(\xi, y)\varphi(y)\, ds(y) \right]$$
$$= \left[l^\pm(\xi) - l(\xi) \right] \varphi(\xi)$$

for almost all $\xi \in \partial S$, where $\Gamma_{\xi,\delta}$ is defined by (1.28).

Proof. Let ξ be a Lebesgue point for φ, and let $|x - \xi| = \delta < \frac{1}{8}r$, with r defined as in (1.18). From (1.15) it follows that $\Gamma_{\xi,8\delta} = \Sigma_1$, so we can write

$$I = \int_{\partial S} k(x, y)\varphi(y)\, ds(y) - \int_{\partial S \setminus \Gamma_{\xi,8\delta}} k(\xi, y)\varphi(y)\, ds(y)$$
$$= I_1 + I_2 + I_3 + I_4,$$

where

$$I_1 = \int_{\Sigma_1} k(x, y)[\varphi(y) - \varphi(\xi)] \, ds(y),$$

$$I_2 = \int_{\Sigma_2} [k(x, y) - k(\xi, y)] [\varphi(y) - \varphi(\xi)] \, ds(y),$$

$$I_3 = \int_{\partial S \setminus \Sigma_{\xi,r}} [k(x, y) - k(\xi, y)] [\varphi(y) - \varphi(\xi)] \, ds(y),$$

$$I_4 = \varphi(\xi) \left[\int_{\partial S} k(x, y) \, ds(y) - \int_{\partial S \setminus \Gamma_{\xi,8\delta}} k(\xi, y) \, ds(y) \right].$$

Let $f(y) = \varphi(y) - \varphi(\xi)$, and let Γ_1 and Γ_2 be defined by (1.17). By Lemmas 1.8 and 2.27,

$$|I_1| \le c_1 \int_{\Sigma_1} |x-y|^{-\gamma} |f(y)| \, ds(y) \le c_2 \int_{\Sigma_1} |x-\xi|^{-\gamma} |f(y)| \, ds(y)$$

$$\le c_3 \delta^{-\gamma} \int_{\Gamma_1} |f(t)| \, dt \to |f(0)| = 0 \quad \text{as} \quad \delta \to 0,$$

where the positive constants c_1, c_2 and c_3 do not depend on x.

By Lemma 1.9,

$$|I_2| \le c_1 \int_{\Sigma_2} |x-\xi| \, |\xi - y|^{-\gamma-1} |f(y)| \, ds(y)$$

$$\le c_4 \delta \int_{\Gamma_2} |s-t|^{-\gamma-1} |f(t)| \, dt,$$

where $c_4 = \text{const} > 0$ does not depend on x. Setting

$$g(t) = \int_s^t |f(\sigma)| \, d\sigma,$$

we see that
$$\frac{1}{s-t}g(t) \to |f(s)| = 0 \quad \text{as} \quad t \to s,$$

g is absolutely continuous [38], and $g' = |f|$ almost everywhere in Γ_2. Let a and b, $a < s < b$, be the arc coordinates of the end-points of $\Sigma_{\xi,r}$. Integrating by parts, we obtain

$$|I_2| \leq c_4 \delta \left[\int_a^{s-8\delta} (s-t)^{-\gamma-1} g'(t)\, dt + \int_{s+8\delta}^b (t-s)^{-\gamma-1} g'(t)\, dt \right]$$

$$= c_5 \left\{ \delta^{-\gamma} [g(s-8\delta) - g(s+8\delta)] \right.$$
$$- \delta [(s-a)^{-\gamma-1} g(a) - (b-s)^{-\gamma-1} g(b)]$$
$$\left. - (\alpha+1)\delta \left[\int_a^{s-8\delta} (s-t)^{-\gamma-2} g(t)\, dt - \int_{s+8\delta}^b (t-s)^{-\gamma-2} g(t)\, dt \right] \right\},$$

where $c_5 = \text{const} > 0$ does not depend on x. By Lemma 2.27, we have $\delta^{-\gamma} g(s \pm 8\delta) \to 0$ as $\delta \to 0$, which also yields

$$\lim_{\delta \to 0} \delta \int_a^{s-8\delta} (s-t)^{-\gamma-2} g(t)\, dt = \lim_{\delta \to 0} \delta \int_{s+8\delta}^b (t-s)g(t)\, dt = 0.$$

Hence, $|I_2| \to 0$ as $\delta \to 0$.

From Lemma 1.10 it follows that

$$|I_3| \leq c_1 \int_{\partial S \setminus \Sigma_{\xi,r}} |x-\xi|\, |\xi - y|^{-\gamma-1} |f(y)|\, ds(y)$$
$$\leq c_6 r^{-2} \delta \int_{\partial S} |f(y)|\, ds(y) \to 0 \quad \text{as} \quad \delta \to 0,$$

where $c_6 = \text{const} > 0$. Finally, as noted in Remark 1.38,

$$\int_{\Sigma_{\xi,8\delta} \setminus \Gamma_{\xi,8\delta}} k(\xi, y) \, ds(y) \to 0 \quad \text{as} \quad \delta \to 0.$$

Consequently, from our assumption on $k(x, y)$ we deduce that

$$I_4 \to [l^{\pm}(\xi) - l(\xi)] \varphi(\xi) \quad \text{as} \quad \delta \to 0,$$

which completes the proof.

2.29. Theorem. *If $k(x, y)$ is a γ-singular kernel on ∂S, $\gamma \in [0, 1)$, and $\varphi \in L^1(\partial S)$, then*

$$\int_{\partial S} k(x, y) \varphi(y) \, ds(y)$$

exists for almost all $x \in \partial S$.

Proof. By Theorem 1.29, the function $\int_{\partial S} |k(x, y)| \, ds(x)$ is continuous on ∂S, therefore, $|\varphi(y)| \int_{\partial S} |k(x, y)| \, ds(x)$ belongs to $L^1(\partial S)$. By Tonelli's Theorem [41], $|k(x, y) \varphi(y)|$ belongs to $L^1(\partial S \times \partial S)$, and the assertion now follows from Fubini's Theorem [41].

2.30. Remark. Using Lebesgue's Dominated Convergence Theorem, it is easy to show that if $\Lambda_{x,\delta}$ is any small neighbourhood of x on ∂S of arc length $\delta > 0$ and $k(x, y)$ and φ are as in Theorem 2.29, then

$$\lim_{\delta \to 0} \int_{\partial S \setminus \Lambda_{x,\delta}} k(x, y) \varphi(y) \, ds(y) = \int_{\partial S} k(x, y) \varphi(y) \, ds(y)$$

for almost all $x \in \partial S$.

2.31. Theorem. *If $\varphi \in L^1(\partial S)$, then*

$$\int_{\partial S} \left[\frac{\partial}{\partial s(y)} \ln|x-y|\right] \varphi(y) \, ds(y)$$

exists in the sense of principal value for almost all $x \in \partial S$.

Proof. Since $\partial \ln|x-y|/\partial \nu(y)$ is 0-singular on ∂S, from Theorem 2.29 it follows that

$$\int_{\partial S} \left[\frac{\partial}{\partial \nu(y)} \ln|x-y|\right] \varphi(y) \, ds(y)$$

exists for almost all $x \in \partial S$. Also, in [29] it is shown that

$$\int_{\partial S} \frac{\varphi(\zeta)}{\zeta - z} \, d\zeta$$

exists in the sense of principal value for almost all $x \in \partial S$. The result is now obtained by means of (1.70).

2.32. Theorem. *If $k(x, y)$ is a proper γ-singular kernel in S_0, $\gamma \in [0, 1)$, $x = \xi \mp \delta\nu(\xi) \in S_0^{\pm}$, $\xi \in \partial S$, $\delta > 0$, and $\varphi \in L^1(\partial S)$, then*

$$\lim_{\delta \to 0} \int_{\partial S} k(x, y)\varphi(y) \, ds(y) = \int_{\partial S} k(\xi, y)\varphi(y) \, ds(y) \qquad (2.65)$$

for almost all $\xi \in \partial S$.

Proof. By Theorem 1.30, $l^{\pm}(\xi)$ and $l(\xi)$ defined by (2.63) and (2.64) are all equal to $\int_{\partial S} k(\xi, y) \, ds(y)$. The desired formula now follows from Theorem 2.28 and Remark 2.30.

Having completed this preparatory work, we can now turn to the study of the behaviour of the elastic potentials with L^2-densities.

2.33. Theorem. *If $\varphi \in L^2(\partial S)$, then*

(i) *$V(\varphi)$ and $W(\varphi)$ are analytic in $\mathbf{R}^2 \setminus \partial S$;*

(ii) *$AV(\varphi) = AW(\varphi) = 0$ in $\mathbf{R}^2 \setminus \partial S$;*

(iii) *Theorem 2.17 holds for $V(\varphi)$ and $W(\varphi)$.*

The proof of this assertion is based on classical results concerning the analyticity of solutions of systems of partial differential equations (see, for example, [26]), the definition of V and W, and their asymptotic expansions for $|x|$ large.

2.34. Theorem. *If $x = \xi \mp \delta \nu(\xi) \in S_0^\pm$, $\xi \in \partial S$, $\delta > 0$, and $\varphi \in L^2(\partial S)$, then*

$$\lim_{\delta \to 0} (V(\varphi))(x) = (V(\varphi))(\xi)$$

for almost all $\xi \in \partial S$.

Proof. From (2.43) and (2.27) we see that the kernel of $V(\varphi)$ is a proper γ-singular kernel in S_0, with any $\gamma \in (0, 1)$. Also, since ∂S is a set of finite measure, we have $\varphi \in L^1(\partial S)$, and the assertion now follows from Theorem 2.32.

2.35. Theorem. *If $x = \xi \mp \delta \nu(\xi) \in S_0^\pm$, $\xi \in \partial S$, $\delta > 0$, and $\varphi \in L^2(\partial S)$, then*

$$\lim_{\delta \to 0} (W(\varphi))(x) = \mp \tfrac{1}{2}\varphi(\xi) + \int_{\partial S} P(\xi, y)\varphi(y)\, ds(y)$$

for almost all $\xi \in \partial S$, where the integral is understood as principal value.

Proof. We examine the terms of the kernel $P(x, y)$ of $W(\varphi)$ one by one, using the expression (2.28). As above, $\varphi \in L^1(\partial S)$.

(i) First, $\partial \ln|x - y|/\partial s(y)$ satisfies the conditions of Theorem 2.28 with $\gamma = 1$ and $l^\pm = l = 0$ (see Theorem 1.58). By Remark 2.30 and

Theorems 2.31 and 2.28,

$$\lim_{\delta\to 0}\int_{\partial S}\left[\frac{\partial}{\partial s(y)}\ln|x-y|\right]\varphi(y)\,ds(y) = \int_{\partial S}\left[\frac{\partial}{\partial s(y)}\ln|\xi-y|\right]\varphi(y)\,ds(y)$$

for almost all $\xi \in \partial S$, where the right-hand side is understood as principal value.

(ii) Next, $\partial \ln|x-y|/\partial \nu(y)$ satisfies the conditions of Theorem 2.28 with $\gamma = 1$, $l^+ = 2\pi$, $l^- = 0$, and $l = \pi$ (see (1.39)–(1.41)). Also, this is a 0-singular kernel on ∂S. Consequently, by Remark 2.30 and Theorem 2.28,

$$\lim_{\delta\to 0}\int_{\partial S}\left[\frac{\partial}{\partial \nu(y)}\ln|x-y|\right]\varphi(y)\,ds(y)$$
$$= \pm\pi\varphi(\xi) + \int_{\partial S}\left[\frac{\partial}{\partial \nu(y)}\ln|\xi-y|\right]\varphi(y)\,ds(y)$$

for almost all $\xi \in \partial S$.

(iii) The kernel $\partial[(x_\alpha - y_\alpha)(x_\beta - y_\beta)|x-y|^{-2}]/\partial s(y)$ satisfies the conditions of Theorem 2.28 with $\gamma = 1$ and $l^\pm = l = 0$, and is 0-singular on ∂S (see Theorem 1.56). Hence, by Remark 2.30 and Theorem 2.28,

$$\lim_{\delta\to 0}\int_{\partial S}\left[\frac{\partial}{\partial s(y)}\frac{(x_\alpha - y_\alpha)(x_\beta - y_\beta)}{|x-y|^2}\right]\varphi(y)\,ds(y)$$
$$= \int_{\partial S}\left[\frac{\partial}{\partial s(y)}\frac{(\xi_\alpha - y_\alpha)(\xi_\beta - y_\beta)}{|\xi-y|^2}\right]\varphi(y)\,ds(y)$$

for almost all $\xi \in \partial S$.

(iv) Finally, the remaining terms are proper γ-singular kernels in S_0, with any $\gamma \in (0,1)$. Therefore, by Theorem 2.32, they satisfy (2.65).

The assertion now follows from (i)–(iv), (2.28) and (2.44).

2.36. Theorem. *If $x = \xi \mp \delta \nu(\xi) \in S_0^\pm$, $\xi \in \partial S$, $\delta > 0$, and $\varphi \in L^2(\partial S)$, then*

$$\lim_{\delta \to 0} (TV(\varphi))(x) = \pm \tfrac{1}{2}\varphi(\xi) + \int_{\partial S} T(\partial_\xi) D(\xi, y) \varphi(y) \, ds(y)$$

for almost all $\xi \in \partial S$, where the integral is understood as principal value.

Proof. From the expressions derived in the proof of Theorem 2.21 it is clear that the kernel of $\partial V_i(\varphi)/\partial x_\alpha$ consists of the same type of terms as that of $W(\varphi)$. To complete the proof, we use (2.54) and (2.56).

2.37. Remark. It can be shown that the same results hold for V, W and TV if $S_0^\pm \ni x \to \xi \in \partial S$ on any direction different from that of $\tau(\xi)$. However, for our purposes it suffices to have the corresponding limiting formulae established as $x \to \xi$ along the normal at ξ.

2.7. Existence of regular solutions

In view of Theorems 2.17 and 2.25 and Remarks 2.22 and 2.24, we may seek the solutions of (N$^+$) and (N$^-$) in the form of $V^+(\varphi)$ and $V^-(\varphi)$ with $\varphi \in C^{0,\alpha}(\partial S)$, that of (D$^+$) in the form of $W^+(\varphi)$ with $\varphi \in C^{1,\alpha}(\partial S)$, and that of (D$^-$) as the sum of $W^-(\varphi)$ with $\varphi \in C^{1,\alpha}(\partial S)$ and a (3×1)-matrix u_0 of the form (2.16). By Theorem 2.20, Corollary 2.23 and Remark 2.24, the boundary value problems (2.39)–(2.42) are reduced respectively to the singular integral eqations

$$-\tfrac{1}{2}\varphi(x) + \int_{\partial S} P(x, y)\varphi(y) \, ds(y) = \mathcal{P}(x), \qquad (\mathcal{D}^+)$$

$$\tfrac{1}{2}\varphi(x) + \int_{\partial S} T(\partial_x) D(x, y)\varphi(y) \, ds(y) = \mathcal{Q}(x), \qquad (\mathcal{N}^+)$$

$$\tfrac{1}{2}\varphi(x) + \int_{\partial S} P(x, y)\varphi(y) \, ds(y) = \mathcal{R}(x) - u_0(x), \qquad (\mathcal{D}^-)$$

$$-\tfrac{1}{2}\varphi(x) + \int_{\partial S} T(\partial_x)D(x,\,y)\varphi(y)\,ds(y) = \mathcal{S}(x), \qquad (\mathcal{N}^-)$$

where $x \in \partial S$ and φ is an unknown density. Let $(\mathcal{D}_0^+)$, $(\mathcal{N}_0^+)$, $(\mathcal{D}_0^-)$, and $(\mathcal{N}_0^-)$ be the corresponding homogeneous equations.

2.38. Theorem. *If $\mathcal{P} \in C^{1,\alpha}(\partial S)$, $\alpha \in (0,1)$, then any solution $\varphi \in C^{0,\alpha}(\partial S)$ of $(\mathcal{D}^+)$ belongs to $C^{1,\alpha}(\partial S)$. A similar statement holds for $(\mathcal{D}^-)$ if $\mathcal{R} \in C^{1,\alpha}(\partial S)$.*

Proof. Using (2.28), (1.70), (1.72), (1.36), (1.66), (1.63), and (1.64), we rewrite $(\mathcal{D}^+)$ in the form

$$(D - \tfrac{1}{2}I)\varphi = \mathcal{P}, \qquad (2.66)$$

where

$$D\varphi = -\frac{1}{2\pi}(\mu'\varepsilon_{\alpha\beta}E_{\alpha\beta}K^s\varphi + K^w\varphi), \qquad (2.67)$$

K^s is defined by (1.72),

$$\begin{aligned} K^w(\varphi) = {} & -(E - i\mu'\varepsilon_{\alpha\beta}E_{\alpha\beta})w(\varphi) - (\lambda' + \mu')\varepsilon_{\alpha\gamma}E_{\gamma\beta}v^e_{\alpha\beta 0}(\varphi) \\ & + \tfrac{1}{2}\varepsilon_{\alpha\beta}(\lambda'E_{3\beta} + h^{-2}E_{\beta 3})v^c_{\alpha 0})(\varphi) - \tfrac{1}{2}E_{3\alpha}v^d_{\alpha 0}(\varphi) \\ & - 2\pi\tilde{W}_0(\varphi), \end{aligned} \qquad (2.68)$$

$$\tilde{W}_0(x) = \int_{\partial S} \tilde{P}(x,\,y)\varphi(y)\,ds(y), \quad x \in \partial S,$$

and $\tilde{P}(x,y)$ satisfies the conditions of Theorem 1.54 with any $\gamma \in (0,1)$. Applying the operator $-2\pi(\mu'\varepsilon_{\alpha\beta}E_{\alpha\beta}K^s - \pi I)$ to both sides of (2.66) and making use of Theorem 1.66, we obtain

$$\bigl[\pi^2(E - \mu'^2 E_{\gamma\gamma})I - \mu'\varepsilon_{\alpha\beta}E_{\alpha\beta}K^s K^w + \pi K^w\bigr]\varphi$$
$$= 2\pi(\mu'\varepsilon_{\alpha\beta}E_{\alpha\beta} - \pi I)\mathcal{P}. \qquad (2.69)$$

Clearly, any solution of (2.66) is also a solution of (2.69). By Theorem 1.65, $C^{1,\alpha}(\partial S)$ is invariant under K^s, consequently, the right-hand side of (2.69) belongs to $C^{1,\alpha}(\partial S)$. By Theorems 1.52, 1.62, 1.61, and 1.54, K^w maps $C^{0,\alpha}(\partial S)$ into $C^{1,\alpha}(\partial S)$. A further application of Theorem 1.65 now shows that every $C^{0,\alpha}(\partial S)$-solution of (2.69) belongs to $C^{1,\alpha}(\partial S)$, which proves the assertion.

The case of $(\mathcal{D}^-)$ is treated similarly.

2.39. Theorem. *The Fredholm Alternative holds for* $(\mathcal{D}^+)$, $(\mathcal{N}^-)$ *and for* $(\mathcal{N}^+)$, $(\mathcal{D}^-)$ *in the (real) dual system* $(C^{0,\alpha}(\partial S), C^{0,\alpha}(\partial S))$, $\alpha \in (0,1)$, *with the bilinear form*

$$(\varphi, \psi) = \int_{\partial S} \varphi^T(y) \psi(y) \, ds(y). \tag{2.70}$$

Proof. Denoting by $\mathcal{D}$ and $\mathcal{N}$ the integral operators occuring in $(\mathcal{D}^\pm)$ and $(\mathcal{N}^\pm)$, respectively, we see that, by (2.25), for any $\varphi, \psi \in C^{0,\alpha}(\partial S)$

$$(\mathcal{D}\varphi, \psi) = \int_{\partial S} \left[\int_{\partial S} P(x,y)\varphi(y) \, ds(y) \right]^T \psi(x) \, ds(x)$$

$$= \int_{\partial S} \left[\int_{\partial S} (T(\partial_y)D(y,x))^T \varphi(y) \, ds(y) \right]^T \psi(x) \, ds(x),$$

$$= \int_{\partial S} \varphi^T(y) \left[\int_{\partial S} T(\partial_y)D(y,x)\psi(x) \, ds(x) \right] ds(y)$$

$$= (\varphi, \mathcal{N}\psi).$$

Due to the symmetry of the bilinear form (2.70), we also have

$$(\mathcal{N}\varphi, \psi) = (\varphi, \mathcal{D}\psi),$$

which means that $\mathcal{D}$ and $\mathcal{N}$ are mutually adjoint in the given dual system.

From (2.67), (2.68), Theorem 1.63, and the fact (pointed out in the proof of Theorem 2.38) that K^w maps $C^{0,\alpha}(\partial S)$ into $C^{1,\alpha}(\partial S)$, it is clear that $C^{0,\alpha}(\partial S)$ is invariant under $\mathcal{D}$.

The kernel k^w of K^w is a proper γ-singular kernel on ∂S with respect to both x and y, for any $\gamma \in (0,1)$. Hence, by Theorem 1.79, K^w is α-regular singular and its complex kernel $\hat{k}^w$ satisfies

$$\hat{k}^w(z, z) = 0, \quad z \in \partial S.$$

Also, (1.72) shows that the same can be said about K^s, except that in this case

$$\hat{k}^s(z, z) = E, \quad z \in \partial S.$$

Then, by (2.67), $\mathcal{D}$ itself is α-regular singular and

$$\hat{k}(z, z) = -\frac{1}{2\pi}[\mu' \varepsilon_{\alpha\beta} E_{\alpha\beta} \hat{k}^s(z, z) + \hat{k}^w(z, z)]$$

$$= -\frac{1}{2\pi}\mu' \varepsilon_{\alpha\beta} E_{\alpha\beta}, \quad z \in \partial S.$$

Consequently, and in view of (2.29),

$$\det\left[-\tfrac{1}{2}E \pm \pi i \hat{k}(z, z)\right] = -\tfrac{1}{8}(1 - \mu'^2) < 0,$$

from which we immediately deduce that the index ϱ of the complex version of $(\mathcal{D}^+)$, defined by (1.86), is zero. According to Theorem 1.82, this means that the Fredholm Alternative holds for $\mathcal{D}$ in the (complex) dual system $(C^{0,\alpha}(\partial S), C^{0,\alpha}(\partial S))$ with the bilinear form

$$(\varphi, \psi) = \int_{\partial S} \varphi^T(\zeta) \psi(\zeta)\, d\zeta,$$

therefore, by Remark 1.83, it also holds for $\mathcal{D}$ in the (real) dual system $(C^{0,\alpha}(\partial S), C^{0,\alpha}(\partial S))$ with the bilinear form (2.70).

The argument is similar for the pair $(\mathcal{D}^-)$, $(\mathcal{N}^+)$.

2.40. Theorem. *The equation $(\mathcal{D}_0^-)$ has precisely three linearly independent $C^{0,\alpha}$-solutions.*

Proof. In view of Theorem 2.38, it suffices to prove the assertion in $C^{1,\alpha}(\partial S)$, $\alpha \in (0,1)$.

It is clear that a (3×1)-matrix u_0 of the form (2.16) is a solution of the interior Neumann problem (N^+). Since $Tu_0 = 0$, replacing u by u_0 in (2.30) we obtain

$$\tfrac{1}{2}u_0(x) + \int_{\partial S} P(x,y)u_0(y)\,ds(y) = 0, \quad x \in \partial S,$$

that is, u_0 is a solution of $(\mathcal{D}_0^-)$. Hence, $\{f^{(1)}, f^{(2)}, f^{(3)}\}$, where

$$\begin{aligned} f^{(1)}(x) &= (1,\,0,\,-x_1)^T, \\ f^{(2)}(x) &= (0,\,1,\,-x_2)^T, \\ f^{(3)}(x) &= (0,\,0,\,1)^T, \end{aligned} \tag{2.71}$$

are three linearly independent solutions of $(\mathcal{D}_0^-)$.

Let $f^{(0)}$ be an arbitrary $C^{1,\alpha}$-solution of $(\mathcal{D}_0^-)$. Then

$$f = f^{(0)} - c_i f^{(i)} \tag{2.72}$$

is also a $C^{1,\alpha}$-solution of $(\mathcal{D}_0^-)$, for any constants c_i. This means that $W^-(f) = 0$ on ∂S, consequently, by Theorems 2.16 and 2.17(i), $W^-(f)$ is a regular solution of the homogeneous exterior Dirichlet problem (D^-). By Theorem 2.15(i), $W^-(f) = 0$ in S^-. This yields $TW^-(f) = 0$ on ∂S, which, in turn, by Theorem 2.25, implies that $TW^+(f) = 0$ on ∂S, and we deduce that $W^+(f)$ is a regular solution of the homogeneous interior

Neumann problem (N^+). Hence, by Theorem 2.15(ii),

$$W^+(f) = W^+(f^{(0)}) - c_i W^+(f^{(i)}) = \tilde{u} \quad \text{in} \quad S^+, \qquad (2.73)$$

where $\tilde{u}$ is of the form (2.16).

Without loss of generality, suppose that the origin of coordinates lies in S^+. We choose the c_i so that $\tilde{u} = 0$, for example, by asking that $(W^+(f))(0) = 0$. This is equivalent to the linear system of equations

$$c_i(W^+(f^{(i)}))(0) = (W^+(f^{(0)}))(0). \qquad (2.74)$$

Let $\{c_1^*, c_2^*, c_3^*\}$ be a solution of the homogeneous system (2.74). Then, setting $f^* = c_i^* f^{(i)}$, we obtain

$$(W^+(f^*))(0) = 0. \qquad (2.75)$$

Taking $f^{(0)} = 0$ and $c_i = c_i^*$ in (2.72), we see that, as above, $W^+(f^*)$ is a regular solution of the homogeneous problem (N^+), therefore, by Theorem 2.15(ii), $W^+(f^*)$ is of the form (2.16). In view of (2.75), we conclude that $W^+(f^*) = 0$ in S^+, so $TW^+(f^*) = 0$ on ∂S. By Theorem 2.25, $TW^-(f^*) = 0$ on ∂S. Thus, $W^-(f^*)$ is a regular solution of the homogeneous exterior Neumann problem (N^-). Hence, by Theorem 2.15(i), $W^-(f^*) = 0$ in S^-. Since $W^+(f^*) = 0$ in S^+, from (2.45) it follows that $f^* = W^-(f^*) - W^+(f^*) = 0$ on ∂S. This means that the homogeneous system (2.74) has only the trivial solution, therefore, (2.74) has a unique solution $\{c_1, c_2, c_3\}$, for which, by (2.73), $W^+(f) = 0$ in S^+. But, as was established earlier, we also have $W^-(f) = 0$ in S^-. Using (2.45) again, we now obtain

$$f = W^-(f) - W^+(f) = 0 \quad \text{on} \quad \partial S.$$

Hence, according to (2.72), any $C^{1,\alpha}$-solution of $(\mathcal{D}_0^-)$ can be expressed uniquely as a linear combination of the $f^{(i)}$.

2.41. Lemma. If $\varphi \in C^{0,\alpha}(\partial S)$, $\alpha \in (0,1)$, is a regular solution of $(\mathcal{N}^-)$, then

$$\int_{\partial S} (\varphi_\alpha - x_\alpha \varphi_3) \, ds = -\int_{\partial S} (\mathcal{S}_\alpha - x_\alpha \mathcal{S}_3) \, ds,$$

$$\int_{\partial S} \varphi_3 \, ds = -\int_{\partial S} \mathcal{S}_3 \, ds.$$

Proof. Setting $u(y) = (c_1, c_2, c - c_1 y_1 - c_2 y_2)^T$ in Theorem 2.9 and taking into account the fact that $Tu = 0$ for such a choice, we find that for $x \in \partial S$

$$\int_{\partial S} [P_{j\alpha}(x,y) - y_\alpha P_{j3}(x,y)] \, ds(y) = -\tfrac{1}{2}(\delta_{j\alpha} - x_\alpha \delta_{j3}),$$

$$\int_{\partial S} P_{j3}(x,y) \, ds(y) = -\tfrac{1}{2}\delta_{j3},$$

or, in view of (2.25),

$$\int_{\partial S} [T_{\alpha k}(\partial_y) - y_\alpha T_{3k}(\partial_y)] D_{kj}(y,x) \, ds(y) = -\tfrac{1}{2}(\delta_{j\alpha} - x_\alpha \delta_{j3}),$$

$$\int_{\partial S} T_{3k}(\partial_y) D_{kj}(y,x) \, ds(y) = -\tfrac{1}{2}\delta_{j3}. \tag{2.76}$$

Multiplying $(\mathcal{N}^-)_3$ and the combinations $(\mathcal{N}^-)_\alpha - x_\alpha \times (\mathcal{N}^-)_3$ by $ds(x)$ and integrating the resulting expressions over ∂S, we obtain the equalities

$$-\tfrac{1}{2}\int_{\partial S} \varphi_3(x) \, ds(x) + \int_{\partial S}\left[\int_{\partial S} T_{3k}(\partial_x) D_{kj}(x,y) \, ds(x)\right] \varphi_j(y) \, ds(y)$$

$$= \int_{\partial S} \mathcal{S}_3(x) \, ds(x)$$

and

$$-\frac{1}{2}\int_{\partial S}[\varphi_\alpha(x) - x_\alpha\varphi_3(x)]\,ds(x)$$

$$-\int_{\partial S}\left\{\int_{\partial S}[T_{\alpha k}(\partial_x)D_{kj}(x,\,y) - x_\alpha T_{3k}(\partial_x)D_{kj}(x,\,y)]\,ds(x)\right\}\varphi_j(y)\,ds(y)$$

$$= \int_{\partial S}[\mathcal{S}_\alpha(x) - x_\alpha\mathcal{S}_3(x)]\,ds(x),$$

and the desired formulae follow from (2.76).

2.42. Theorem. (i) *The interior Dirichlet problem* (D$^+$) *has a unique regular solution for any* $\mathcal{P} \in C^{1,\alpha}(\partial S)$, $\alpha \in (0,1)$. *This solution can be represented as the extension* $W^+(\varphi)$ *to* $\bar{S}^+$ *of the restriction to* S^+ *of a double layer potential* $W(\varphi)$ *with density* $\varphi \in C^{1,\alpha}(\partial S)$.

(ii) *The exterior Neumann problem* (N$^-$) *has a unique regular solution for any* $\mathcal{S} \in C^{0,\alpha}(\partial S)$, $\alpha \in (0,1)$, *if and only if*

$$\begin{aligned}\int_{\partial S}(\mathcal{S}_a - x_\alpha \mathcal{S}_3)\,ds &= 0, \\ \int_{\partial S}\mathcal{S}_3\,ds &= 0.\end{aligned} \quad (2.77)$$

This solution can be represented as the restriction $V^-(\varphi)$ *to* $\bar{S}^-$ *of a single layer potential* $V(\varphi)$ *with density* $\varphi \in C^{0,\alpha}(\partial S)$.

Proof. By Theorem 2.39, the Fredholm Alternative holds for (D$^+$), (N$^-$) and (D$^-$), (N$^+$) in the (real) dual system $(C^{0,\alpha}(\partial S), C^{0,\alpha}(\partial S))$ with the bilinear form (2.70).

Let u be a regular solution of (N$^-$), and consider a disk Γ_R of sufficiently large radius R so that $\bar{S}^+ \subset \Gamma_R$. Applying Theorem 2.5 in $S^- \cap \Gamma_R$, we find that

$$\int_{\partial S}(S_a - x_\alpha S_3)\,ds = \int_{\partial S}(T_{\alpha i} - x_\alpha T_{3i})u_i\,ds = \int_{\partial \Gamma_R}(T_{\alpha i} - x_\alpha T_{3i})u_i\,ds,$$

$$\int_{\partial S} S_3\,ds = \int_{\partial S} T_{3i}u_i\,ds = \int_{\partial \Gamma_R} T_{3i}u_i\,ds,$$

from which (2.77) are obtained by letting $R \to \infty$ and taking (2.38) into account.

Suppose now that (2.77) hold, and let $\varphi^{(0)}$ be a regular solution of $(\mathcal{N}_0^-)$. By (2.55) and (2.57), this is equivalent to $TV^-(\varphi^{(0)}) = 0$ on ∂S. Since $AV^-(\varphi^{(0)}) = 0$ in S^- and, by Lemma 2.41 and Theorem 2.17(ii), $V^-(\varphi^{(0)}) \in \mathcal{A}$, it follows that $V^-(\varphi^{(0)})$ is a solution of the homogeneous exterior Dirichlet problem (D^-). By Theorem 2.15(i), $V^-(\varphi^{(0)}) = 0$ in S^-, hence, by Theorem 2.19, $V^-(\varphi^{(0)}) = 0 = V^+(\varphi^{(0)})$ on ∂S.

Now $V^+(\varphi^{(0)})$ is a solution of the homogeneous interior Dirichlet problem (D^+), consequently, by Theorem 2.15(i), $V^+(\varphi^{(0)}) = 0$ in S^+. Then $TV^+(\varphi^{(0)}) = 0$ on ∂S, and (2.57) and (2.55) yield

$$\varphi^{(0)} = TV^+(\varphi^{(0)}) - TV^-(\varphi^{(0)}) = 0 \quad \text{on} \quad \partial S,$$

from which we conclude that $(\mathcal{N}_0^-)$ has only the zero solution. According to the Fredholm Alternative, so does (D_0^+), therefore, (D^+) and $(\mathcal{N}^-)$ have unique solutions $\varphi \in C^{0,\alpha}(\partial S)$.

To complete the proof, we remark that in the case of (N^-) it follows from Lemma 2.41, (2.77) and Theorem 2.17(ii) that $V^-(\varphi) \in \mathcal{A}$, in other words, $V^-(\varphi)$ is a regular solution of (N^-). At the same time, in the case of (D^+) Theorem 2.38 yields $\varphi \in C^{1,\alpha}(\partial S)$, hence, by Theorem 2.25, $W^+(\varphi)$ is a regular solution of the problem.

The uniqueness of these solutions was established in Theorem 2.15(i).

2.43. Theorem. *The interior Neumann problem* (N$^+$) *is soluble for any* $\mathcal{Q} \in C^{0,\alpha}(\partial S)$, $\alpha \in (0,1)$, *if and only if*

$$\int_{\partial S} (\mathcal{Q}_\alpha - x_\alpha \mathcal{Q}_3) \, ds = 0,$$
$$\int_{\partial S} \mathcal{Q}_3 \, ds = 0. \tag{2.78}$$

The regular solution is unique up to a (3×1)-*matrix of the form* (2.16) *and can be represented as the restriction* $V^+(\varphi)$ *to* $\bar{S}^+$ *of a single layer potential* $V(\varphi)$ *with density* $\varphi \in C^{1,\alpha}(\partial S)$.

Proof. By Theorem 2.39 and the Fredholm Alternative, ($\mathcal{N}^+$) is soluble if and only if

$$(f^{(i)}, \mathcal{Q}) = \int_{\partial S} (f^{(i)})^T \mathcal{Q} \, ds = 0,$$

where the $f^{(i)}$ are defined by (2.71). Expliciting these conditions, we see that they coincide with (2.78). Consequently, if the equalities (2.78) hold, then there is a density $\varphi \in C^{0,\alpha}(\partial S)$ for which $V^+(\varphi)$ is a regular solution of (N$^+$). The uniqueness of this solution is discussed in Theorem 2.15(ii).

2.44. Theorem. *The exterior Dirichlet problem* (D$^-$) *has a unique regular solution for any* $\mathcal{R} \in C^{1,\alpha}(\partial S)$. *This solution can be represented as the sum of the extension* $W^-(\varphi)$ *to* $\bar{S}^-$ *of the restriction to* S^- *of a double layer potential* $W(\varphi)$ *with density* $\varphi \in C^{1,\alpha}(\partial S)$ *and a particular* (3×1)-*matrix* u_0 *of the form* (2.16).

Proof. According to Theorem 2.39 and the Fredholm Alternative, ($\mathcal{N}_0^+$) has precisely three linearly independent $C^{0,\alpha}$-solutions $g^{(i)}$. Without loss of generality, suppose that the sets $\{f^{(i)}\}$ and $\{g^{(i)}\}$ have been biorthonormalized [24], that is, we have

$$(f^{(i)}, g^{(j)}) = \delta_{ij}.$$

Taking $u_0 = c_i f^{(i)}$, where $c_i = \int_{\partial S} (g^{(i)})^T \mathcal{R} \, ds$, we see that

$$(g^{(j)}, \mathcal{R} - c_i f^{(i)}) = \int_{\partial S} (g^{(j)})^T (\mathcal{R} - c_i f^{(i)}) \, ds = 0.$$

Consequently, by the Fredholm Alternative, $(\mathcal{D}^-)$ has a unique solution $\varphi \in C^{0,\alpha}(\partial S)$. By Theorem 2.38, $\varphi \in C^{1,\alpha}(\partial S)$. Since, according to Theorem 2.17(i), $W^-(\varphi) + u_0 \in \mathcal{A}^*$, it follows that $W^-(\varphi) + u_0$ is a regular solution of $(\mathcal{D}^-)$. The uniqueness of this solution is guaranteed by Theorem 2.15(i).

2.45. Remark. The restrictions (2.78) and (2.77), which are necessary and sufficient for the solubility of (N^+) and (N^-), respectively, have a direct physical meaning. By Remark 2.1, they represent the condition that the transverse shear force and the bending and twisting moments acting on ∂S should be zero.

The regular solutions to all our boundary value problems have been found in closed form. But one question still remains unanswered: what is the mechanical significance of the class $\mathcal{A}$ that intervenes so essentially in the proceedings? Is its introduction really necessary? Could there be regular solutions outside this class as well? The boundary integral equation method, while elegant and precise, offers no answer. To settle this outstanding matter, in Chapter 3 we change over to a different technique of investigation, equally powerful, which allows us to obtain the complete integral of the system (2.37).

2.8. Smoothness of the integrable solutions

We conclude this chapter by taking a closer look at the regularity properties of the L^2-solutions of the singular integral equations corresponding to the interior and exterior Dirichlet and Neumann boundary value problems.

2.46. Theorem. *Suppose that*

$$\lambda\varphi(x) + \int_{\partial S} k(x, y)\varphi(y)\,ds(y) = f(x) \tag{2.79}$$

for almost all $x \in \partial S$, where $k(x, y)$ is a proper γ-singular kernel on ∂S, $\gamma \in [0, 1)$, $\lambda \in \mathbf{R}$, $\lambda \neq 0$, and $f \in C^{0,\alpha}(\partial S)$, $\alpha \in (0, 1]$. If $\varphi \in L^p(\partial S)$ is a solution of (2.66), then $\varphi \in C^{0,\beta}(\partial S)$, with $\beta = \min\{\alpha, 1 - \gamma\}$ if $\gamma \in (0, 1)$, $\beta = \alpha$ if $\alpha \in (0, 1)$ and $\gamma = 0$, and any $\beta \in (0, 1)$ if $\alpha = 1$ and $\gamma = 0$.

Proof. Let

$$K(x) = \int_{\partial S} k(x, y)\varphi(y)\,ds(y),$$

which, by Theorem 2.29, exists for almost all $x \in \partial S$. We have

$$|K(x)| \leq \int_{\partial S} \big[|k(x,y)|^{2-\gamma}|\varphi(y)|^p\big]^{1/[p(2-\gamma)]}$$
$$\times |\varphi(y)|^{(1-\gamma)/(2-\gamma)}|k(x,y)|^{(p-1)/p}\,ds(y).$$

Setting

$$p_1 = p(2 - \gamma), \quad p_2 = \frac{p(2-\gamma)}{1-\gamma}, \quad p_3 = \frac{p}{p-1}$$

and noting that

$$\frac{1}{p_1} + \frac{1}{p_2} + \frac{1}{p_3} = 1$$

and that the three factors of the integrand on the right-hand side above belong to $L^{p_1}(\partial S)$, $L^{p_2}(\partial S)$ and $L^{p_3}(\partial S)$, respectively, we apply the generalized Hölder inequality and Theorem 1.29 to obtain

$$|K(x)| \le \left[\int_{\partial S} |k(x,y)|^{2-\gamma} |\varphi(y)|^p \, ds(y)\right]^{1/[p(2-\gamma)]}$$

$$\times \left[\int_{\partial S} |\varphi(y)|^p \, ds(y)\right]^{(1-\gamma)[p(2-\gamma)]} \left[\int_{\partial S} |k(x,y)| \, ds(y)\right]^{(p-1)/p}$$

$$= c_1 \|\varphi\|_p^{(1-\gamma)/(2-\gamma)} \left[\int_{\partial S} |k(x,y)|^{2-\gamma} |\varphi(y)|^p \, ds(y)\right]^{1/[p(2-\gamma)]},$$

where $c_1 = \text{const} > 0$. Then, by Fubini's Theorem,

$$\int_{\partial S} |K(x)|^{p(2-\gamma)} \, ds(x)$$

$$\le c_1 \|\varphi\|_p^{p(1-\gamma)} \int_{\partial S} \left[\int_{\partial S} |k(x,y)|^{2-\gamma} |\varphi(y)|^p \, ds(y)\right] ds(x)$$

$$= c_1 \|\varphi\|_p^{p(1-\gamma)} \int_{\partial S} \left[\int_{\partial S} |k(x,y)|^{2-\gamma} \, ds(x)\right] |\varphi(y)|^p \, ds(y).$$

Since

$$|k(x,y)|^{2-\gamma} \le c_2 |x-y|^{-\gamma(2-\gamma)}, \quad c_2 = \text{const} > 0,$$

and $0 \le \gamma(2-\gamma) < 1$, from Theorem 1.29 it follows that

$$\int_{\partial S} |K(x,y)|^{p(2-\gamma)} \, ds(y) \le c_3 \|\varphi\|_p^{p(1-\gamma)} \int_{\partial S} |\varphi(y)|^p \, ds(y) = c_3 \|\varphi\|_p^{p(2-\gamma)},$$

where $c_3 = \text{const} > 0$. This means that $K \in L^{p(2-\gamma)}(\partial S)$. Then (2.79) yields $\varphi \in L^{p(2-\gamma)}(\partial S)$. Applying the argument successively n times, we deduce that $\varphi \in L^{p(2-\gamma)^n}(\partial S)$ for any positive integer n. Hence, $\varphi \in L^\infty(\partial S)$.

If we now repeat the proof of Theorem 1.30 with the integrals understood in the sense of Lebesgue, we conclude that $K \in C^{0,\delta}(\partial S)$, with $\delta = 1 - \gamma$ for $\gamma \in (0,1)$ and any $\delta \in (0,1)$ for $\gamma = 0$. The result now follows from (2.79).

2.47. Theorem. *Suppose that the equations* $(\mathcal{D}^\pm)$ *and* $(\mathcal{N}^\pm)$ *hold almost everywhere on* ∂S, *and that* $\mathcal{P}, \mathcal{Q}, \mathcal{R}, \mathcal{S} \in C^{0,\alpha}(\partial S)$, $\alpha \in (0,1)$. *If* $\varphi \in L^2(\partial S)$ *is a solution of any of the above equations, then* $\varphi \in C^{0,\alpha}(\partial S)$.

Proof. $(\mathcal{D}^\pm)$ and $(\mathcal{N}^\pm)$ are of the form

$$(K - \omega I)\varphi = g, \tag{2.80}$$

where, as seen in the proof of Theorem 2.39, K is α-regular singular and $\omega \in \mathbf{R}$, $\omega \neq 0$. In [27] it is shown that we can always find an α-regular singular operator L that maps $L^2(\partial S)$ into $L^2(\partial S)$, and a $\vartheta \in \mathbf{R}$, $\vartheta \neq 0$, such that the equation

$$(L - \vartheta I)(K - \omega I)\varphi = (L - \vartheta I)g \tag{2.81}$$

is of the form (2.79), where $\lambda \in \mathbf{R}$, $\lambda \neq 0$, $f \in C^{0,\alpha}(\partial S)$, and $k(x,y)$ is a proper $(1-\alpha)$-singular kernel on ∂S. Since every solution of (2.80) is also a solution of (2.81), the result follows from Theorem 2.46 with $p = 2$.

3 Complex variable treatment

3.1. Complex representation of the stresses

We revert to the original notation, where S is a bounded simply connected domain in $\mathbf{R}^2$, whose boundary ∂S is a closed simple contour.

In agreement with (2.37), we consider the homogeneous system (2.3), that is,
$$N_{\alpha\beta,\beta} - N_{3\alpha} = 0, \qquad (3.1)$$
$$N_{3\beta,\beta} = 0,$$

and investigate its analytic solutions in S.

From $(3.1)_2$ we deduce that there is a function $\mathcal{G}(x_\gamma)$ such that
$$N_{31} = \mathcal{G}_{,2}, \qquad (3.2)$$
$$N_{32} = -\mathcal{G}_{,1}.$$

This and $(3.1)_1$ yield
$$N_{11,1} + (N_{12} - \mathcal{G})_{,2} = 0,$$
$$(N_{12} + \mathcal{G})_{,1} + N_{22,2} = 0.$$

Hence, there are functions $\mathcal{H}_\alpha(x_\gamma)$ such that
$$N_{11} = \mathcal{H}_{1,2}, \qquad N_{12} - \mathcal{G} = -\mathcal{H}_{1,1}, \qquad (3.3)$$
$$N_{22} = -\mathcal{H}_{2,1}, \qquad N_{12} + \mathcal{G} = \mathcal{H}_{2,2}.$$

Obviously, we must have
$$\mathcal{G} - \mathcal{H}_{1,1} = -\mathcal{G} + \mathcal{H}_{2,2}. \qquad (3.4)$$

Let $\mathcal{B}(x_\gamma)$ be such that
$$\mathcal{G} = \mathcal{B}_{,12} . \tag{3.5}$$

Then from (3.4) it follows that there is a function $\mathcal{C}(x_\gamma)$ satisfying
$$\mathcal{B}_{,2} - \mathcal{H}_1 = -\mathcal{C}_{,2} ,$$
$$\mathcal{B}_{,1} - \mathcal{H}_2 = \mathcal{C}_{,1} ,$$

in which case (3.2), (3.3) and (3.5) imply that
$$\begin{aligned} N_{11} &= (\mathcal{C} + \mathcal{B})_{,22} , \\ N_{22} &= (\mathcal{C} - \mathcal{B})_{,11} , \\ N_{12} &= -\mathcal{C}_{,12} , \\ N_{31} &= \mathcal{B}_{,122} , \\ N_{32} &= -\mathcal{B}_{,112} . \end{aligned} \tag{3.6}$$

The stress functions $\mathcal{B}$ and $\mathcal{C}$ generate a deformation state if and only if $N_{\alpha\beta}$ and $N_{3\alpha}$ given by (3.6) satisfy the compatibility relations (2.7). Replacing (3.6) in (2.7), we obtain the Cauchy-Riemann system

$$(h^2 \Delta \mathcal{B}_{,12} - \mathcal{B}_{,12})_{,1} = (1 - \sigma)(\Delta \mathcal{C} + \mathcal{B}_{,22} - \mathcal{B}_{,11})_{,2} ,$$
$$(h^2 \Delta \mathcal{B}_{,12} - \mathcal{B}_{,12})_{,2} = -(1 - \sigma)(\Delta \mathcal{C} + \mathcal{B}_{,22} - \mathcal{B}_{,11})_{,1} .$$

Hence,
$$\Delta \mathcal{B}_{,12} - \frac{1}{h^2} \mathcal{B}_{,12} = \frac{2}{h^2} \operatorname{Re} \Omega_0, \tag{3.7}$$

$$\Delta \mathcal{C} + \mathcal{B}_{,22} - \mathcal{B}_{,11} = \frac{2}{1 - \sigma} \operatorname{Im} \Omega_0, \tag{3.8}$$

where Ω_0 is an arbitrary analytic function of $z = x_1 + ix_2$ in S.

Let $\eta(z,\bar{z})$ be an arbitrary real solution in S of the equation

$$\Delta\eta - \frac{1}{h^2}\eta = 0. \tag{3.9}$$

Then from (3.7) we find that

$$\mathcal{B}_{,12} = \mathrm{Re}\left[\eta(z,\bar{z}) - 2\Omega_0(z)\right]. \tag{3.10}$$

For simplicity, in what follows we omit the explicit mention of z and $\bar{z}$ in the symbols of functions.

Applying the operator $(\Delta - h^{-2})\partial^2/\partial x_1 \partial x_2$ to (3.8) and using (3.10), we obtain

$$\Delta\left(\Delta\mathcal{C}_{,12} - \frac{1}{h^2}\mathcal{C}_{,12}\right) = \frac{8}{h^2}\sigma_1 \,\mathrm{Re}\,\Omega_0,$$

where $\sigma_1 = \frac{1}{4}(1-2\sigma)(1-\sigma)^{-1}$. Therefore,

$$\mathcal{C}_{,12} = \mathrm{Re}\,(\theta - 2\sigma_1\bar{z}\Omega_0' + 2w_0), \tag{3.11}$$

where θ is another arbitrary real solution of (3.9) in S and w_0 an arbitrary analytic function in S. From (3.10) and (3.11) we deduce that

$$(\mathcal{C}+\mathcal{B})_{,12} = \mathrm{Re}\,(\eta + \theta - 2\Omega_0 + 2w_0 - 2\sigma_1\bar{z}\Omega_0'), \tag{3.12}$$
$$(\mathcal{C}-\mathcal{B})_{,12} = \mathrm{Re}\,(-\eta + \theta + 2\Omega_0 + 2w_0 - 2\sigma_1\bar{z}\Omega_0'). \tag{3.13}$$

Differentiating (3.12) with respect to x_2 and replacing the result in (3.8) differentiated with respect to x_1, and dealing similarly with (3.13) and (3.7), we find $(\mathcal{C}-\mathcal{B})_{,111}$ and $(\mathcal{C}+\mathcal{B})_{,222}$, which we then combine with $(\mathcal{C}-\mathcal{B})_{,122}$ and $(\mathcal{C}+\mathcal{B})_{,112}$ obtained directly from (3.12) and (3.13). Thus, we arrive at

$$\Delta\left[(\mathcal{C}+\mathcal{B})_{,2} + i(\mathcal{C}-\mathcal{B})_{,1}\right] = 2(\eta_{,z} - 4\sigma_1\Omega_0'),$$

where $(\ldots)_{,z} = \partial(\ldots)/\partial z$. This implies that

$$(C+B)_{,2} = 2\operatorname{Re}(2h^2\eta_{,z} - \sigma_1\bar{z}\Omega_0 + \Omega_1),$$
$$(C-B)_{,1} = 2\operatorname{Im}(2h^2\eta_{,z} - \sigma_1\bar{z}\Omega_0 - \Omega_2),$$
(3.14)

where Ω_α are arbitrary analytic functions in S.

Since this representation has been obtained by differentiating the exact formula (3.8), it may contain too much arbitrariness. Replacing (3.14) in (3.12), (3.13) and (3.8), we see that

$$\operatorname{Re}(4h^2\eta_{,zz} - \theta + \Omega_1 - \Omega_2' - 2\omega_0) = 0,\qquad(3.15)$$

$$2(1-\sigma)(\Omega_1' + \Omega_2') + (3-2\sigma)\Omega_0 = 0.\qquad(3.16)$$

Now (3.16) yields
$$\Omega_0 = -\frac{2(1-\sigma)}{3-2\sigma}(\Omega_1' + \Omega_2').\qquad(3.17)$$

Since η and θ are solutions of (3.9), from (3.15) we find that

$$2\operatorname{Re}\omega_0 = \operatorname{Re}(\Omega_1' - \Omega_2'),$$
$$\theta = 4h^2\operatorname{Re}\varphi_{,zz}.$$
(3.18)

Substituting (3.17) and (3.18) in (3.10)–(3.12) and (3.14) and setting

$$\kappa = \frac{1-2\sigma}{3-2\sigma} = \frac{\mu}{2\lambda + 3\mu},\qquad(3.19)$$

we obtain

$$(C+B)_{,2} = \operatorname{Re}\left[4h^2\eta_{,z} + \kappa\bar{z}(\Omega_1' + \Omega_2') + 2\Omega_1\right],$$
$$(C-B)_{,1} = \operatorname{Im}\left[4h^2\eta_{,z} + \kappa\bar{z}(\Omega_1' + \Omega_2') - 2\Omega_2\right],$$
$$C_{,12} = \operatorname{Re}\left[4h^2\eta_{,zz} + \kappa\bar{z}(\Omega_1'' + \Omega_2'') + \Omega_1' - \Omega_2'\right],$$
$$B_{,12} = \operatorname{Re}\left[\eta + (1+\kappa)(\Omega_1' + \Omega_2')\right].$$
(3.20)

Finally, from the above relations and the formulae (3.6) we conclude that

$$N_{11} = -\operatorname{Im}\left[4h^2\eta_{,zz} + \kappa\bar{z}(\Omega_1'' + \Omega_2'') - \kappa\Omega_2' + (2-\kappa)\Omega_1'\right],$$
$$N_{22} = \operatorname{Im}\left[4h^2\eta_{,zz} + \kappa\bar{z}(\Omega_1'' + \Omega_2'') + \kappa\Omega_1' - (2-\kappa)\Omega_2'\right],$$
$$N_{12} = -\operatorname{Re}\left[4h^2\eta_{,zz} + \kappa\bar{z}(\Omega_1'' + \Omega_2'') + \Omega_1' - \Omega_2'\right], \quad (3.21)$$
$$N_{31} = -\operatorname{Im}\left[2\eta_{,z} + (1+\kappa)(\Omega_1'' + \Omega_2'')\right],$$
$$N_{32} = -\operatorname{Re}\left[2\eta_{,z} + (1+\kappa)(\Omega_1'' + \Omega_2'')\right].$$

Since an arbitrary solution of (3.9) can be expressed in terms of an arbitrary analytic function in S [26], the bending and twisting moments and the transverse shear forces are represented in terms of three arbitrary analytic functions of z in S. Functions of this type are known in the literature as complex potentials.

3.2. The traction boundary value problem

We consider the Neumann boundary conditions

$$N_i = N_{i\alpha}\nu_\alpha = \tilde{N}_i \quad \text{on} \quad \partial S. \quad (3.22)$$

According to Remark 2.1, the resultant force and complex moment acting on an arc $t_0 t$ of ∂S are

$$[\mathcal{N}]_{t_0}^t = \int_{t_0}^t N_3\, ds = \tilde{\mathcal{N}},$$
$$[\mathcal{M}]_{t_0}^t = \int_{t_0}^t \left[-N_2 + x_2 N_3 + i(N_1 - x_1 N_3)\right] ds = \tilde{\mathcal{M}}. \quad (3.23)$$

From (3.6) and (3.22) we obtain

$$N_3 = \frac{d}{ds}\mathcal{B}_{,12},$$
$$-N_2 + x_2 N_3 + i(N_1 - x_1 N_3) \qquad (3.24)$$
$$= \frac{d}{ds}\left[(\mathcal{C} - \mathcal{B})_{,1} + i(\mathcal{C} + \mathcal{B})_{,2} - iz\mathcal{B}_{,12}\right].$$

Using this in (3.23), we find that on ∂S

$$\begin{aligned}\mathcal{B}_{,12} &= \tilde{\mathcal{N}} + \beta_1, \quad \beta_1 \in \mathbf{R},\\ (\mathcal{C} - \mathcal{B})_{,1} + i(\mathcal{C} + \mathcal{B})_{,2} - iz\mathcal{B}_{,12} &= \tilde{\mathcal{M}} + \beta_2, \quad \beta_2 \in \mathbf{C}.\end{aligned} \qquad (3.25)$$

Setting $\Omega_1 + \Omega_2 = \varrho$ and $\Omega_1 - \Omega_2 = \vartheta$, from (3.20) and (3.25) we now deduce that

$$\begin{aligned} 4h^2 \eta_{,\bar{z}} + \kappa z \bar{\varrho}' + \varrho + \bar{\vartheta} &= z\tilde{\mathcal{N}} - i\tilde{\mathcal{M}} + \beta_1 z - i\beta_2,\\ \eta + \tfrac{1}{2}(1+\kappa)(\varrho' + \bar{\varrho}') &= \tilde{\mathcal{N}} + \beta_1 \quad \text{on} \quad \partial S. \end{aligned} \qquad (3.26)$$

Hence, the traction boundary value problem reduces to finding η, ϱ and ϑ satisfying (3.9) in S and (3.26) on ∂S.

3.3. The displacement boundary value problem

Consider the Dirichlet boundary conditions

$$u_i = \tilde{u}_i \quad \text{on} \quad \partial S. \qquad (3.27)$$

We introduce the complex displacements, moments and force by

$$\begin{aligned}\Gamma &= u_1 + iu_2,\\ \Theta &= u_3,\\ \Phi &= N_{11} - N_{22} + 2iN_{12}, \qquad (3.28)\\ \Psi &= N_{11} + N_{22},\\ \Lambda &= N_{31} + iN_{32}.\end{aligned}$$

Then the constitutive relations (2.5) become

$$\Phi = 4h^2 \mu \Gamma_{,\bar{z}},$$
$$\Psi = 4h^2(\lambda + \mu)\operatorname{Re}\Gamma_{,z}, \qquad (3.29)$$
$$\Lambda = \mu(\Gamma + 2\Theta_{,\bar{z}}),$$

and from (3.21), (3.28)$_3$ and (3.29)$_1$ we find that

$$\Gamma = -\frac{i}{2h^2\mu}\left[4h^2\eta_{,\bar{z}} + \kappa z(\Omega_1' + \Omega_2') + \bar{\Omega}_1 - \bar{\Omega}_2 + \theta_1\right], \qquad (3.30)$$

where θ_1 is an analytic function of z in S. This, (3.21) and (3.29) yield

$$\operatorname{Im}\theta_1' = -\kappa \operatorname{Im}(\Omega_1' + \Omega_2'),$$

therefore,

$$\theta_1 = -\kappa(\Omega_1 + \Omega_2) - c_1 z - c_2, \qquad (3.31)$$

where $c_1 \in \mathbf{R}$ and $c_2 \in \mathbf{C}$.

Setting $\Omega_\alpha = w_\alpha'$, from (3.30) and (3.31) we deduce that

$$\Gamma = -\frac{i}{2h^2\mu}\left[4h^2\eta_{,\bar{z}} + \kappa z(\bar{w}_1'' + \bar{w}_2'') - \kappa(w_1' + w_2') + \bar{w}_1' - \bar{w}_2' - c_1 z - c_2\right]. \qquad (3.32)$$

From this, (3.21) and (3.29)$_3$ we obtain

$$\Theta = \frac{i}{4h^2\mu}\left[\kappa z(\bar{w}_1' + \bar{w}_2') - \kappa \bar{z}(w_1' + w_2') + \bar{w}_1 - \bar{w}_2\right.$$
$$\left. - c_1 z\bar{z} - c_2 \bar{z} - 2h^2(1+\kappa)(\bar{w}_1'' + \bar{w}_2'') + \theta_2\right], \qquad (3.33)$$

where θ_2 is an analytic function of z in S. Since Θ is real, we must have

$$c_1 = 0,$$
$$\theta_2 = -(w_1 + w_2) + 2h^2(1+\kappa)(w_1'' + w_2'') + \bar{c}_2 z - ic_3, \qquad (3.34)$$

where $c_3 \in \mathbf{R}$. We set

$$\psi = -\frac{2}{\mu}\eta,$$

$$\Omega = \frac{i\kappa}{2h^2\mu}(w'_1 + w'_2),$$

$$\omega = \frac{i}{2h^2\mu}(w_1 - w_2), \tag{3.35}$$

$$l = l_1 + il_2 = \frac{ic_2}{2h^2\mu},$$

$$m = \frac{c_3}{4h^2\mu}.$$

From this and (3.32)–(3.34) we then conclude that

$$\begin{aligned}\Gamma &= i\psi_{,\bar{z}} + z\bar{\Omega}' + \Omega + \bar{\omega}' + l, \\ \Theta &= \operatorname{Re}\left(4h^2\frac{\lambda + 2\mu}{\mu}\Omega' - \bar{z}\Omega - \omega - l\bar{z} + m\right).\end{aligned} \tag{3.36}$$

3.1. Remark. The matrix u_0 defined by the terms containing l and m is of the form (2.16), consequently, it represents a rigid displacement. These terms are unessential and in what follows we assume them to be incorporated in Ω and ω, respectively.

Let $z \in \partial S$. Writing $\tilde{u} = \tilde{u}_1 + i\tilde{u}_2$, from (3.27) we find that

$$\begin{aligned}i\psi_{,\bar{z}} + z\bar{\Omega}' + \Omega + \bar{\omega}' &= \tilde{u}, \\ \operatorname{Re}\left(4h^2\frac{\lambda + 2\mu}{\mu}\Omega' - \bar{z}\Omega - \omega\right) &= \tilde{u}_3.\end{aligned} \tag{3.37}$$

Thus, the displacement boundary value problem reduces to finding a solution ψ in S of (3.9) and arbitrary analytic functions Ω and ω in S satisfying (3.37) on ∂S.

From (3.29), (3.35) and (3.36) we see that

$$\Phi = 4h^2\mu(i\psi_{,\bar{z}\bar{z}} + z\bar{\Omega}'' + \bar{\omega}''),$$
$$\Psi = 4h^2(\lambda + \mu)(\Omega' + \bar{\Omega}'), \qquad (3.38)$$
$$\Lambda = i\mu\psi_{,\bar{z}} + 4h^2(\lambda + 2\mu)\bar{\Omega}''.$$

Comparing the definitions of η, ϱ, ϑ and ψ, Ω, ω, we can rewrite the traction boundary conditions (3.26) as

$$i\psi_{,\bar{z}} + z\bar{\Omega}' - \frac{2\lambda + 3\mu}{\mu}\Omega + \bar{\omega}' = -\frac{1}{2h^2\mu}(\tilde{\mathcal{M}} + iz\tilde{\mathcal{N}} + i\beta_1 z + \beta_2),$$
$$i\psi - 4h^2\frac{\lambda + 2\mu}{\mu}(\Omega' - \bar{\Omega}') = -\frac{2i}{\mu}(\tilde{\mathcal{N}} + \beta_1), \qquad (3.39)$$

and, using (3.23), (3.24) and (3.20), the resultant force and complex moment acting on the arc $t_0 t$ of ∂S as

$$[\mathcal{N}]_{t_0}^t = \left[-\tfrac{1}{2}\mu\psi - 2 1h^2(\lambda + 2\mu)(\Omega' - \bar{\Omega}')\right]_{t_0}^t,$$
$$[\mathcal{M}]_{t_0}^t = \left[\tfrac{1}{2}i\mu z\psi - 2h^2(\lambda + 2\mu)z(\Omega - \bar{\Omega}') \right. \qquad (3.40)$$
$$\left. - 2h^2\mu(i\psi_{,\bar{z}} + z\bar{\Omega}' + \bar{\omega}') + 2h^2(2\lambda + 3\mu)\Omega\right]_{t_0}^t.$$

3.2. Remark. A representation similar to (3.38) has been derived in the case of Reissner's theory [21]. In our notation this is

$$\Phi = 4h^2\mu\left[i\varpi_{,\bar{z}\bar{z}} + \frac{\lambda + 2\mu}{4(\lambda + \mu)}(z\bar{\Omega}'' + \bar{\omega}'')\right],$$
$$\Psi = \frac{h^2\mu(3\lambda + 2\mu)}{\lambda + \mu}(\Omega' + \bar{\Omega}'),$$
$$\Lambda = \mu(\tfrac{5}{6}i\varpi_{,\bar{z}} + 4h^2\bar{\Omega}''),$$

where ϖ is an arbitrary analytic solution in S of the equation

$$\Delta\varpi - \frac{5}{6h^2}\varpi = 0.$$

3.3. Remark. Θ given by $(3.36)_2$ satisfies $\Delta\Delta\Theta = 0$. From (3.36) and (3.19) we obtain

$$\Gamma = -2\Theta_{,\bar{z}} + i\psi_{,\bar{z}} + 4h^2\mu^{-1}(\lambda+2\mu)\bar{\Omega}''.$$

Hence, in this theory, as in Kirchhoff's, $\Theta = u_3$ remains a biharmonic function. In addition, Kirchhoff's theory also leads to $(1.5)_2$, that is, $\Gamma = -2\Theta_{,\bar{z}}$. Here Γ contains two correction terms, of which one is a solution of (3.9) and the other is harmonic.

3.4. Remark. If instead of (2.5) we adopt Mindlin's constitutive relations [25], then, ignoring rigid displacements, we obtain

$$\Gamma = i\chi_{,\bar{z}} - 2k^2\left(\bar{z}\bar{\Omega}' + \Omega + \bar{\omega}' + \frac{8h^2}{1-\sigma}\bar{\Omega}''\right),$$

$$\Theta = 2\operatorname{Re}(\bar{z}\Omega + \omega),$$

$$\Phi = \frac{2Eh^2}{1+\sigma}\left[i\chi_{,\bar{z}\bar{z}} - 2k^2\left(\bar{z}\bar{\Omega}'' + \bar{\omega}'' + \frac{8h^2}{1-\sigma}\bar{\Omega}'''\right)\right],$$

$$\Psi = -8k^2\frac{Eh^2}{1-\sigma}\operatorname{Re}\Omega',$$

$$\Lambda = k^2\frac{E}{2(1-\sigma)}\left[i\chi_{,\bar{z}} + (k^2-1)(\bar{z}\bar{\Omega}' - \Omega + \bar{\omega}') + \frac{8h^2}{1-\sigma}\bar{\Omega}''\right],$$

where E is Young's modulus, k^2 a correction coefficient introduced by Mindlin, and χ an arbitrary real solution of

$$\Delta\chi - \frac{k^2}{h^2}\chi = 0.$$

3.4. Arbitrariness in the complex potentials

Suppose that the functions ψ_*, Ω_* and ω_* generate the same stress state as ψ, Ω and ω. Then from $(3.38)_2$ we find that

$$\operatorname{Re} \Omega'_* = \operatorname{Re} \Omega',$$

therefore,

$$\Omega_* = \Omega + id_1 z + d_2, \quad d_1 \in \mathbf{R}, \quad d_2 \in \mathbf{C}. \tag{3.41}$$

Using $(3.38)_3$ and (3.41), we obtain

$$\psi_{*,\bar{z}} = \psi_{,\bar{z}},$$

which yields

$$\psi_* = \psi + d_3 z + d_4, \quad d_3, d_4 \in \mathbf{C}.$$

Since ψ and ψ_* are real functions, it follows that $d_3 = 0$ and $d_4 \in \mathbf{R}$, and the fact that both ψ and ψ_* are solutions of (3.9) leads to

$$\psi_* = \psi. \tag{3.42}$$

From $(3.38)_1$, (3.41) and (3.42) we deduce that

$$\omega_* = \omega + d_5 z + id_6, \quad d_5, d_6 \in \mathbf{C}. \tag{3.43}$$

Choosing d_1, d_2 and d_5 so that

$$d_1 = \frac{\kappa}{2h^2\mu(1+\kappa)}\beta_1, \quad d_2 - \kappa \bar{d}_5 = \frac{\kappa}{2h^2\mu}\beta_2,$$

we make the terms β_1 and $i\beta_1 z + \beta_2$ vanish in (3.39). To reduce the arbitrariness of Ω and ω we may impose, for example, the additional conditions

$$\Omega(0) = 0 \quad (\text{or } \omega'(0) = 0), \quad \omega(0) = 0.$$

If we also want the displacements to remain unchanged, then, according to (3.36) and (3.41)–(3.43), we must require that

$$d_5 = -\bar{d}_2, \quad d_6 \in \mathbf{R}.$$

Thus, the functions Ω and ω are completely determined if we ask, say, that they should satisfy

$$\Omega(0) = 0, \quad (\text{or } \omega'(0) = 0), \quad \text{Im } \Omega'(0) = 0, \quad \omega(0) = 0.$$

3.5. Bounded multiply connected domain

Let S be multiply connected. For the sake of simplicity, we introduce the notation

$$\theta(z, \bar{z}) \in \mathcal{U} \quad \Leftrightarrow \quad \theta(z, \bar{z}) \text{ is single-valued in } S.$$

From (3.36) and (3.38) we see that the moments, forces and displacements are single-valued if

$$i\psi_{,z} + z\bar{\Omega}' + \Omega + \bar{\omega}' \in \mathcal{U},$$

$$\text{Re}\left(4h^2 \frac{\lambda + 2\mu}{\mu} \Omega' - \bar{z}\Omega - \omega\right) \in \mathcal{U},$$

$$i\psi_{,z\bar{z}} + z\bar{\Omega}'' + \bar{\omega}'' \in \mathcal{U} \tag{3.44}$$

$$\text{Re } \Omega' \in \mathcal{U},$$

$$i\psi_{,z} + 4h^2 \frac{\lambda + 2\mu}{\mu} \bar{\Omega}'' \in \mathcal{U}.$$

Clearly, from $(3.44)_1$ it follows that

$$\Omega'' \in \mathcal{U}, \tag{3.45}$$

which, in view of $(14.1)_5$, yields $\psi_{,z} \in \mathcal{U}$. Then $\psi_{,z\bar{z}} \in \mathcal{U}$ also, consequently,

$$\psi \in \mathcal{U}. \tag{3.46}$$

From $(3.44)_3$, (3.45) and (3.46) we deduce that

$$\omega'' \in \mathcal{U}, \tag{3.47}$$

and from $(3.44)_{2,4}$

$$\mathrm{Re}\,(\bar{z}\Omega + \omega) \in \mathcal{U}. \tag{3.48}$$

Also, $(3.44)_1$ and (3.46) lead to

$$z\bar{\Omega}' + \Omega + \bar{\omega}' \in \mathcal{U}. \tag{3.49}$$

This means that the necessary single-valuedness conditions are $(3.44)_4$ and (3.45)–(3.49).

Suppose now that the boundary of S consists of $n+1$ disjoint simple closed curves of which one, ∂S, encloses all the remaining ones, ∂S_k, $k = 1,\ldots,n$. According to a well-known argument in three-dimensional elasticity [28], we can choose arbitrary points z_k inside the contours ∂S_k and write

$$\begin{aligned}\Omega &= \frac{1}{2\pi i} \sum_{k=1}^{n} (c_k z + d_k) \log(z - z_k) + \tilde{\Omega}, \\ \omega &= \frac{1}{2\pi i} \sum_{k=1}^{n} (p_k z + q_k) \log(z - z_k) + \tilde{\omega},\end{aligned} \tag{3.50}$$

where $c_k, d_k, p_k, q_k \in \mathbf{C}$, $k = 1,\ldots,n$, and $\tilde{\Omega}$ and $\tilde{\omega}$ are analytic functions in S. From $(3.44)_4$ and (3.48)–(3.50) we find that the coefficients must satisfy

$$\mathrm{Re}\,c_k = 0, \qquad d_k + \bar{p}_k = 0, \qquad \mathrm{Re}\,q_k = 0. \tag{3.51}$$

Traversing ∂S_k once anticlockwise, from (3.40), (3.46) and (3.50) we obtain the resultant force and moment on ∂S_k in the form

$$\mathcal{N}_k = -[\mathcal{N}]_{\partial S_k} = -4h^2(\lambda + 2\mu)\,\mathrm{Im}\, c_k,$$
$$\mathcal{M}_k = -[\mathcal{M}]_{\partial S_k} = 2h^2\left[-(2\lambda + 3\mu)d_k + \mu\bar{p}_k\right].$$

Combining these relations with (3.51), we deduce that

$$c_k = -2\pi i c \mathcal{N}_k, \qquad d_k = -2\pi c \mathcal{M}_k,$$
$$p_k = 2\pi c \bar{\mathcal{M}}_k, \qquad q_k = -2\pi i c s_k,$$

where $c = \left[8\pi h^2(\lambda + 2\mu)\right]^{-1}$ and $s_k \in \mathbf{R}$. From this and (14.7) we now conclude that

$$\begin{aligned}\Omega &= -c\sum_{k=1}^{n}(z\mathcal{N}_k - i\mathcal{M}_k)\log(z - z_k) + \tilde{\Omega}, \\ \omega &= -c\sum_{k=1}^{n}(iz\bar{\mathcal{M}}_k + s_k)\log(z - z_k) + \tilde{\omega}.\end{aligned} \qquad (3.52)$$

3.5. Remark. The terms $s_k \log(z - z_k)$, although many-valued, do not alter the single-valuedness of the force, moments and displacements. These terms occur only in the expression of Θ, in the form $\mathrm{Re}\left[s_k \log(z - z_k)\right] = s_k \log|z - z_k| \in \mathcal{U}$.

3.6. Unbounded multiply connected domain

Suppose that the curve ∂S has expanded to infinity. Introducing the notation

$$\mathrm{N} = \sum_{k=1}^{n}\mathcal{N}_k, \qquad \mathrm{M} = \sum_{k=1}^{n}\mathcal{M}_k, \qquad s = \sum_{k=1}^{n}s_k$$

and proceeding as in [28], from (3.52) we find that the complex potentials admit the expansions

$$\Omega = -c(N - iM)\log z + \sum_{n=-\infty}^{\infty} a_n z^n,$$

$$\omega = -c(i\bar{M}z + s)\log z + \sum_{n=-\infty}^{\infty} b_n z^n,$$

(3.53)

where $a_n, b_n \in \mathbf{C}$. Then (3.38) and (3.53) yield the complex moments and force in the form

$$\Phi = 4h^2\mu\bigg[i\psi_{,\bar{z}\bar{z}} - cNz\bar{z}^{-1} + ic\bar{M}z\bar{z}^{-2} + icM\bar{z}^{-1} + cs\bar{z}^{-2}$$

$$+ \sum_{n=-\infty}^{\infty} n(n-1)(\bar{a}_n z + \bar{b}_n)\bar{z}^{n-2}\bigg],$$

$$\Psi = 4h^2(\lambda + \mu)\bigg[-2cN\ln|z| - 2cN + ic(Mz^{-1} - \bar{M}\bar{z}^{-1})$$

$$+ \sum_{n=-\infty}^{\infty} n(a_n z^{n-1} + \bar{a}_n \bar{z}^{n-1})\bigg],$$

(3.54)

$$\Lambda = i\mu\psi_{,z} + 4h^2(\lambda + 2\mu)\bigg[-cN\bar{z}^{-1} + ic\bar{M}\bar{z}^{-2}$$

$$+ \sum_{n=-\infty}^{\infty} n(n-1)\bar{a}_n \bar{z}^{n-2}\bigg].$$

To investigate the behaviour of Φ, Ψ and Λ as $|z| \to \infty$ we need to know the asymptotics of ψ. Since [1]

$$\frac{d}{d\xi}K_0(\xi) = -K_1(\xi),$$

$$\frac{d}{d\xi}K_1(\xi) = -K_0(\xi) - \frac{1}{\xi}K_1(\xi)$$

(3.55)

and, as $|\xi| \to \infty$,

$$K_0(\xi) = \left(\frac{\pi}{2\xi}\right)^{\frac{1}{2}} e^{-\xi} + \cdots,$$
$$K_1(\xi) = \left(\frac{\pi}{2\xi}\right)^{\frac{1}{2}} e^{-\xi} + \cdots,$$
(3.56)

and since $(2\pi)^{-1} K_0(h^{-1}|x-y|)$ is a fundamental solution of (3.9), just as in harmonic potential theory we deduce that ψ admits the representation

$$\psi(x) = \int_{\cup \partial S_k} \left[\psi(y) \frac{\partial}{\partial \nu(y)} K_0(h^{-1}|x-y|) \right.$$
$$\left. - K_0(h^{-1}|x-y|) \frac{\partial}{\partial \nu(y)} \psi(y) \right] ds(y).$$

From this, (3.55) and (3.56) we see that ψ and its derivatives vanish as $|x| \to \infty$. Therefore, (3.54) shows that Φ, Ψ and Λ are bounded at infinity if and only if

$$N = 0, \quad a_n = 0 \ (n \geq 2), \quad b_n = 0 \ (n \geq 3). \tag{3.57}$$

Next, using (3.53) and (3.57) in (3.36), we obtain

$$\Gamma = (a_1 + \bar{a}_1)z + 2\bar{b}_2\bar{z} + 2ic\mathrm{M}\ln|z| - ic\bar{\mathrm{M}}z\bar{z}^{-1} + a_0 + \bar{b}_1 + ic\mathrm{M}$$
$$+ a_{-1}z^{-1} - cs\bar{z}^{-1} - \bar{a}_{-1}z\bar{z}^{-2} + O(|z|^{-2}),$$
$$\Theta = \mathrm{Re}\left[ic(\bar{\mathrm{M}}z - \mathrm{M}\bar{z})\log z + cs\ln|z| - b_2 z^2 - a_1 z\bar{z}\right. \tag{3.58}$$
$$\left. - (\bar{a}_0 + b_1)z + 4h^2 \frac{\lambda + 2\mu}{\mu} a_1 - a_{-1}z^{-1}\bar{z} - b_0\right]$$
$$+ O(|z|^{-1}).$$

Since $u_3 = \Theta$ occurs in the internal energy density (2.14) only in terms of its derivatives, we conclude that for a finite energy solution, that is,

$\Gamma = O(1)$ and $\Theta = O(\ln |z|)$ as $|z| \to \infty$, we must have

$$M = 0, \quad a_1 = i\alpha \quad (\alpha \in \mathbf{R}), \quad b_2 = 0, \quad b_1 = -\bar{a}_0. \tag{3.59}$$

In view of (3.41), we may discard the a_1-term. Setting $-cs = a \in \mathbf{R}$ and $\operatorname{Re} b_0 = b$, from (3.53), (3.57) and (3.59) we conclude that, as $|z| \to \infty$,

$$\begin{aligned} \Omega &= \sum_{n=-\infty}^{-1} a_n z^n, \\ \omega &= a \log z + b + \sum_{n=-\infty}^{-1} b_n z^n. \end{aligned} \tag{3.60}$$

These formulae and (3.36) (with an arbitrary ψ satisfying (3.9)) yield the general analytic finite energy solution of (2.37) in S.

3.6. Remark. It is interesting to note that, although ψ is responsible for the sixth order character of this bending theory, it plays no active role in the far-field pattern of the solution, which depends exclusively on the structure of Ω and ω.

3.7. Remark. A straightforward calculation shows that for $a = b$ in (3.60) the expansion of u coincides with (2.36), which characterizes the class $\mathcal{A}$. Also, from (3.36), (3.56) and (3.60) we obtain the asymptotics

$$\Phi = O(|z|^{-2}), \quad \Psi = O(|z|^{-2}), \quad \Lambda = O(|z|^{-3}),$$
$$\Gamma = O(|z|^{-1}), \quad \Theta = O(\ln |z|).$$

These imply that the Betti formula in the exterior domain, proved in Theorem 2.12, holds for all solutions satisfying (3.57) and (3.59). Consequently, the condition that $u \in \mathcal{A}$, which was shown to be sufficient for the solvability of the exterior Neumann problem in the case $n = 1$, turns out to be also necessary if we want a unique solution. Removing the restriction

$a = b$ in (3.60) means that the regular solution of this problem is unique up to an arbitrary vertical translation.

3.7. Example

Consider an infinite plate with a circular hole of radius ρ, acted upon at the hole by a normal force cx_3, $c = \text{const} > 0$, parallel to the middle plane of the plate. Choosing the origin at the centre of the hole and following the averaging procedure set out in §2.1, we arrive at the boundary and far-field conditions

$$N_{rr} = h^2 c, \quad N_{r\theta} = N_{3r} = 0 \quad \text{if} \quad |z| = \rho,$$
$$N_{\alpha\beta} = N_{3\alpha} = 0 \quad \text{as} \quad |z| \to \infty, \tag{3.61}$$

where N_{rr}, $N_{r\theta}$ and N_{3r} are the physical polar components of the $N_{i\alpha}$, defined by

$$N_{rr} = \tfrac{1}{2} \operatorname{Re}(e^{-2i\theta}\Phi + \Psi),$$
$$N_{r\theta} = \tfrac{1}{2} \operatorname{Im}(e^{-2i\theta}\Phi - \Psi),$$
$$N_{3r} = \operatorname{Re}(e^{-i\theta}\Lambda).$$

To solve the problem we use a semi-inverse method, setting

$$a_{-1} = a_{-2} = \ldots = 0,$$
$$b_{-1} = b_{-2} = \ldots = 0,$$
$$\psi = 0,$$
$$\operatorname{Im} b = 0$$

in (3.60), the last value being justified by the arbitrariness in ω as shown by (3.43). Then from (3.60) and (3.38) we see that (3.61) are satisfied if

$a = -c\rho^2(2\mu)^{-1}$, in which case (3.36) and (3.60) yield

$$\Gamma = -\frac{c\rho^2}{2\mu}\bar{z}^{-1},$$

$$\Theta = \frac{c\rho^2}{2\mu}\ln|z| - b.$$

It is clear that this exterior Neumann problem has a unique solution in $\mathcal{A}$, corresponding to $b = -c\rho^2(2\mu)^{-1}$. If this restriction is removed, then the solution is determined up to an arbitrary vertical translation, as noted in Remark 3.7.

3.8. Physical significance of the restrictions

From (3.54) and (3.57) we find that, as $|z| \to \infty$, the limiting values of Φ, Ψ and Λ are

$$\Phi_\infty = 8h^2 \mu \bar{b}_2,$$
$$\Psi_\infty = 8h^2(\lambda + \mu)\operatorname{Re} a_1,$$
$$\Lambda_\infty = 0,$$

that is, the bending and twisting moments are uniformly distributed at infinity, while the transverse shear force vanishes. We can see that $(3.59)_{2,3}$ are equivalent to $\Phi_\infty = \Psi_\infty = 0$.

In view of (1.6), the rotations in the vertical coordinate planes in $\mathbf{R}^3$ are given by

$$\varepsilon_\alpha = \tfrac{1}{2}(u_\alpha - u_{3,\alpha}).$$

From this, (3.28), (3.58) and $(3.59)_{2,3}$ we deduce that, as $|z| \to \infty$, the complex vertical rotation is given by

$$\varepsilon = \varepsilon_1 + i\varepsilon_2 = \tfrac{1}{2}(\Gamma - 2\Theta_{,\bar{z}})$$
$$= 2ic M \ln|z| - ic\bar{M}z\bar{z}^{-1} + a_0 + \bar{b}_1 + icM + O(|z|^{-1}).$$

Hence, $(3.59)_{1,4}$ are equivalent to $\varepsilon_\infty = 0$. In this case $\Theta = O(\ln|z|)$.

3.8. Remark. In view of the above arguments, we conclude that an analytic solution of (2.37) is of finite energy if and only if the corresponding bending and twisting moments, transverse shear force and rotation in the vertical coordinate planes vanish at infinity. Then, by Remark 3.7, $\mathcal{A}$ is the class of all finite energy solutions of (2.37) that contain no vertical translation.

4 Generalized Fourier series

4.1. The interior Dirichlet problem

In this chapter we suspend the convention of summation over repeated indices, as well as that regarding the values taken by Latin subscripts. Greek subscripts and superscripts continue to take the values 1, 2.

4.1. Definition. Let X be a normed space. A subset $\mathcal{X} \subset X$ is called a *fundamental set in* X if span $\mathcal{X}$ is dense in X.

The following assertion is a well-known result of functional analysis.

4.2. Theorem. *If X is a Hilbert space, then $\mathcal{X} \subset X$ is a fundamental set in X if and only if the orthogonal complement of $\mathcal{X}$ in X consists o the zero vector alone.*

Let ∂S_* be a simple closed C^2-curve such that ∂S lies strictly in the domain S_*^+ enclosed by ∂S_*, and let $\{x^{(k)} \in \partial S_*, \ k = 1, 2, \ldots\}$ be a countable set of points densely distributed on ∂S_*. We set $S_*^- = \mathbf{R}^2 \setminus \bar{S}_*^+$, and denote by $D^{(i)}$ the columns of the matrix D.

4.3. Theorem. *The set*

$$\{f^{(i)}, \ \theta^{(jk)}, \ i, j = 1, 2, 3, \ k = 1, 2, \ldots\}, \tag{4.1}$$

where the $f^{(i)}$ are defined by (2.71) and

$$\theta^{(jk)}(x) = D^{(j)}(x, x^{(k)}), \tag{4.2}$$

is linearly independent on ∂S and fundamental in $L^2(\partial S)$.

Proof. Suppose that there are a positive integer N and real numbers c_i and c_{jk}, $i, j = 1, 2, 3$, $k = 1, 2, \ldots, N$, not all zero, such that

$$\sum_{i=1}^{3} c_i f^{(i)}(x) + \sum_{j=1}^{3} \sum_{k=1}^{N} c_{jk} \theta^{(jk)}(x) = 0, \quad x \in \partial S. \tag{4.3}$$

Setting

$$\varpi(x) = \sum_{i=1}^{3} f^{(i)}(x) + \sum_{j=1}^{3} \sum_{k=1}^{N} c_{jk} \theta^{(jk)}(x), \tag{4.4}$$

from (4.2), (4.3) and Theorem 2.8 we see that

$$A(\partial_x)\varpi(x) = 0, \quad x \in S^+,$$
$$\varpi(x) = 0, \quad x \in \partial S,$$

that is, ϖ is a regular solution of the homogeneous interior Dirichlet problem. By Theorem 2.15(i), $\varpi = 0$ in $\bar{S}^+$. Then, using analyticity arguments, we deduce that

$$\varpi(x) = 0, \quad x \in S^+_*. \tag{4.5}$$

Let $x^{(p)}$ be any of the (finitely many) points $x^{(1)}, \ldots, x^{(N)}$. In view of (4.4) and (4.2), we write

$$\varpi_l(x) = \sum_{i=1}^{3} c_i f_l^{(i)}(x) + \sum_{j=1}^{3} \sum_{k=1}^{N} c_{jk} D_{jl}(x, x^{(k)})$$

and remark that, according to (2.27), as $x \to x^{(p)}$ all the terms on the right-hand side remain bounded except $c_{lp} D_{ll}(x, x^{(p)})$, which is of order $O(\ln |x - x^{(p)}|)$. This clearly contradicts the equality (4.5), and we conclude that all the c_{jk} in (4.3) must be zero. Since the $f^{(i)}$ are linearly independent, we deduce that the c_i are also zero. Hence, the set (4.1) is linearly independent on ∂S.

Now let $\varphi \in L^2(\partial S)$ be such that for all $i, j = 1, 2, 3$ and $k = 1, 2, \ldots$

$$\int_{\partial S} (f^{(i)})^T \varphi \, ds = \int_{\partial S} (\theta^{(jk)})^T \varphi \, ds = 0. \tag{4.6}$$

By (4.2) and (2.24), this is equivalent to

$$\int_{\partial S} D(x^{(k)}, y) \varphi(y) \, ds(y) = 0, \quad k = 1, 2, \ldots, \tag{4.7}$$

$$\int_{\partial S} [\varphi_\alpha(y) - y_\alpha \varphi_3(y)] \, ds(y) = \int_{\partial S} \varphi_3(y) \, ds(y) = 0. \tag{4.8}$$

Consider the elastic single layer potential of density φ

$$V(x) = \int_{\partial S} D(x, y) \varphi(y) \, ds(y).$$

Since, by Theorem 2.33(i), V is continuous on ∂S_* and the points $x^{(k)}$, $k = 1, 2, \ldots$, are densely distributed on ∂S_*, from (4.7) it follows that $V = 0$ on ∂S_*. In view of Theorem 2.33(ii, iii), we have

$$A(\partial_x)V(x) = 0, \quad x \in S_*^-,$$

$$V(x) = 0, \quad x \in \partial S_*,$$

$$V \in \mathcal{A}.$$

This means that V is a regular solution in $\bar{S}_*^-$ of the homogeneous exterior Dirichlet problem (D$^-$), consequently, by Theorem 2.15(i), $V = 0$ in $\bar{S}_*^-$. The analyticity of the elastic single layer potential V in $\mathbf{R}^2 \setminus \partial S$ now implies that

$$V(x) = 0, \quad x \in S^-. \tag{4.9}$$

In turn, this yields $TV = 0$ in S^-. Letting $S^- \ni x' \to x \in \partial S$ along the support line of $\nu(x)$, from Theorem 2.36 we find that

$$-\tfrac{1}{2}\varphi(x) + \int_{\partial S} T(\partial_x) D(x,y) \varphi(y)\, ds(y) = 0$$

for almost all $x \in \partial S$, where the integral is understood as principal value. By Theorem 2.47, $\varphi \in C^{0,\alpha}(\partial S)$ with any $\alpha \in (0,1)$. Then V is continuous in $\mathbf{R}^2$ and

$$A(\partial_x) V(x) = 0, \quad x \in S^+,$$

$$V(x) = 0, \quad x \in \partial S,$$

that is, V is a regular solution in $\bar{S}^+$ of the homogeneous problem (D^+). Consequently, by Theorem 2.15(i), $V = 0$ in $\bar{S}^+$. From this and (4.9) we deduce that $(TV)^+ = (TV)^- = 0$ on ∂S, and (2.55) yields $\varphi = 0$.

Since $L^2(\partial S)$ is a Hilbert space, we now apply Theorem 4.2 to conclude that (4.1) is a fundamental set in $L^2(\partial S)$.

Let u be the (unique) regular solution of (D^+). By Theorem 2.9 and (2.39),

$$u(x) = \int_{\partial S} D(x,y) \psi(y)\, ds(y) - F(x), \quad x \in S^+, \tag{4.10}$$

$$F(x) = \int_{\partial S} D(x,y) \psi(y)\, ds(y), \quad x \in S^-, \tag{4.11}$$

where we have used the notation

$$F(x) = \int_{\partial S} P(x,y) \mathcal{P}(y)\, ds(y), \quad x \in \mathbf{R}^2 \setminus \partial S, \tag{4.12}$$

$$\psi(y) = (Tu)(y), \quad y \in \partial S. \tag{4.13}$$

147

The formula (4.11) yields

$$\int_{\partial S} D(x^{(k)}, y)\psi(y)\, ds(y) = F(x^{(k)}), \quad k = 1, 2, \ldots,$$

which, by (2.24) and (4.2), is equivalent to

$$\int_{\partial S} (\theta^{(jk)})^{\mathrm{T}} \psi\, ds = F_j(x^{(k)}), \quad j = 1, 2, 3, \ k = 1, 2, \ldots \quad (4.14)$$

We arrange the elements of (4.1) in the order

$$f^{(1)}, f^{(2)}, f^{(3)}, \theta^{(11)}, \theta^{(21)}, \theta^{(31)}, \ldots, \theta^{(1k)}, \theta^{(2k)}, \theta^{(3k)}, \ldots,$$

and denote the new sequence by $\{\theta^{(m)}\}_{m=1}^{\infty}$. Let $\{\omega^{(n)}\}_{n=1}^{\infty}$ be the orthonormalized fundamental sequence constructed from the set $\{\theta^{(m)}\}_{m=1}^{\infty}$ in $L^2(\partial S)$ by means of the Gram-Schmidt process. Then

$$\omega^{(n)} = \sum_{m=1}^{n} k_{nm} \theta^{(m)}, \quad n = 1, 2, \ldots,$$

where k_{nm} are well-determined numbers. Writing

$$\psi^{(n)} = \sum_{r=1}^{n} p_r \omega^{(r)}, \quad n = 1, 2, \ldots, \quad (4.15)$$

with the coefficients on the right-hand side given by

$$p_r = \int_{\partial S} (\omega^{(r)})^{\mathrm{T}} \psi\, ds = \sum_{m=1}^{r} k_{rm} \int_{\partial S} (\theta^{(m)})^{\mathrm{T}} \psi\, ds, \quad r = 1, 2, \ldots, \quad (4.16)$$

and setting

$$u^{(n)}(x) = \int_{\partial S} D(x, y) \psi^{(n)}(y)\, ds(y) - F(x), \quad x \in S^+, \quad (4.17)$$

from (4.10) we see that for $x \in S^+$

$$|u(x) - u^{(n)}(x)| \leq \sum_{i=1}^{3} |u_i(x) - u_i^{(n)}(x)|$$

$$\leq \sum_{i=1}^{3} \int_{\partial S} |(D^{(i)}(y, x))^T [\psi(y) - \psi^{(n)}(y)]| \, ds(y)$$

$$\leq \sum_{i=1}^{3} \|D^{(i)}(x, \cdot)\|_2 \|\psi - \psi^{(n)}\|_2.$$

Since the $\|D^{(i)}(x, \cdot)\|_2$ are uniformly bounded on any closed subdomain $S' \subset S^+$ and $\|\psi - \psi^{(n)}\|_2 \to 0$ as $n \to \infty$, we conclude that $u^{(n)} \to u$ uniformly on S'.

Clearly, each $u^{(n)}$ is a solution of the equation $Au = 0$ in S^+.

4.4. Remark. According to (4.13), u is a regular solution of the interior Neumann problem

$$A(\partial_x)u(x) = 0, \quad x \in S^+,$$
$$T(\partial_x)u(x) = \psi(x), \quad x \in \partial S,$$

therefore, by Theorem 2.43,

$$\int_{\partial S} (f^{(i)})^T \psi \, ds = 0, \quad i = 1, 2, 3, \tag{4.18}$$

which is equivalent to (2.78). Since $\theta^{(i)} = f^{(i)}$, $i = 1, 2, 3$, from (4.16) and (4.18) it now follows that

$$p_1 = p_2 = p_3 = 0,$$
$$p_r = \sum_{m=4}^{r} k_{rm} \int_{\partial S} (\theta^{(m)})^T \psi \, ds, \quad r = 4, 5, \ldots \tag{4.19}$$

Hence, the approximate solution $u^{(n)}$ is given by (4.17), where $F(x)$ and $\psi^{(n)}$ are given by (4.12) and (4.15), respectively, with the p_r as in (4.19) and fully determined for $r = 4, 5, \ldots$ by (4.14) and (4.12).

4.2. The interior Neumann problem

With the notation introduced in the preceding section we can prove the following assertion.

4.5. Theorem. *The set*

$$\{f^{(i)}, \vartheta^{(jk)}, \ i, j = 1, 2, 3, \ k = 1, 2, \ldots\}, \qquad (4.20)$$

where the $f^{(i)}$ are defined by (2.71) and

$$\vartheta^{(jk)}(x) = T(\partial_x) D^{(j)}(x, x^{(k)}), \qquad (4.21)$$

is linearly independent on ∂S and fundamental in $L^2(\partial S)$.

Proof. As in the proof of Theorem 4.3, suppose that there are a positive integer N and real numbers c_i and c_{jk}, $i, j = 1, 2, 3$, $k = 1, 2, \ldots, N$, not all zero, such that

$$\sum_{i=1}^{3} c_i f^{(i)}(x) + \sum_{j=1}^{3} \sum_{k=1}^{N} c_{jk} \vartheta^{(jk)}(x) = 0, \quad x \in \partial S. \qquad (4.22)$$

Then, taking (4.21), (4.22) and Theorem 2.8 into consideration, we find that the (3×1)-matrix

$$\varpi(x) = \sum_{j=1}^{3} \sum_{k=1}^{N} c_{jk} D^{(j)}(x, x^{(k)}) \qquad (4.23)$$

is a regular solution of the interior Neumann problem

$$A(\partial_x)\varpi(x) = 0, \quad x \in S^+,$$

$$T(\partial_x)\varpi(x) = -\sum_{i=1}^{3} c_i f^{(i)}(x), \quad x \in \partial S,$$

consequently, by (2.78),

$$\int_{\partial S} (f^{(l)})^{\mathrm{T}} \left[-\sum_{i=1}^{3} c_i f^{(i)} \right] ds = 0, \quad l = 1, 2, 3,$$

which implies that the coefficients c_1, c_2 and c_3 are all equal to zero. This yields

$$T(\partial_x)\varpi(x) = 0, \quad x \in \partial S,$$

so, by Theorem 2.15(ii),

$$\varpi(x) = \sum_{i=1}^{3} \beta_i f^{(i)}(x), \quad x \in \bar{S}^+,$$

for some constants β_i, $i = 1, 2, 3$. From this and (4.23) it follows that

$$\tilde{\varpi}(x) = \sum_{j=1}^{3}\sum_{k=1}^{N} c_{jk} D^{(j)}(x, x^{(k)}) - \sum_{i=1}^{3} \beta_i f^{(i)}(x) = 0, \quad x \in \bar{S}^+.$$

By analyticity, $\tilde{\varpi} = 0$ in $\bar{S}^+_*$, and the linear independence of the set (4.20) on ∂S is established by means of the argument used in the proof of Theorem 4.3.

Suppose now that for all $i, j = 1, 2, 3$ and $k = 1, 2, \ldots$ the function $\varphi \in L^2(\partial S)$ satisfies

$$\int_{\partial S} (f^{(i)})^{\mathrm{T}} \varphi \, ds = \int_{\partial S} (\vartheta^{(jk)})^{\mathrm{T}} \varphi \, ds = 0.$$

According to (4.21) and (2.25), this means that

$$\int_{\partial S} [\varphi_\alpha(y) - y_\alpha \varphi_3(y)] \, ds(y) = \int_{\partial S} \varphi_3(y) \, ds(y) = 0, \qquad (4.24)$$

$$\int_{\partial S} P(x^{(k)}, y) \varphi(y) \, ds(y) = 0, \quad k = 1, 2, \ldots \qquad (4.25)$$

By Theorem 2.33(i), the double layer potential of density φ

$$W(x) = \int_{\partial S} P(x, y) \varphi(y) \, ds(y)$$

is continuous on ∂S_*. Since the $x^{(k)}$ are densely distributed on ∂S_*, from (4.25) we deduce that $W = 0$ on ∂S_*. Then, by Theorem 2.33(ii, iii), W is a regular solution of the exterior Dirichlet problem

$$A(\partial_x) W(x) = 0, \quad x \in S_*^-,$$
$$W(x) = 0, \quad x \in \partial S_*,$$
$$W \in \mathcal{A},$$

hence, by Theorem 2.15(i), $W = 0$ in $\bar{S}_*^-$. The analyticity of W in $\mathbf{R}^2 \setminus \partial S$ now yields $W = 0$ in S^-. Letting $S^- \ni x' \to x \in \partial S$ along the support line of $\nu(x)$, from Theorem 2.35 we find that

$$\tfrac{1}{2} \varphi(x) + \int_{\partial S} P(x, y) \varphi(y) \, ds(y) = 0$$

for almost all $x \in \partial S$, where the integral is understood as principal value. By Theorem 2.47, $\varphi \in C^{0,\alpha}(\partial S)$ for any $\alpha \in (0,1)$, which, in view of Theorem 2.40, implies that

$$\varphi(x) = \sum_{i=1}^{3} \gamma_i f^{(i)}(x), \quad x \in \partial S, \quad \gamma_i = \text{const} > 0.$$

Then, by (4.24),

$$\int_{\partial S} \left[(f^{(l)})^T \sum_{i=1}^{3} \gamma_i f^{(i)} \right] ds = 0, \quad l = 1, 2, 3,$$

and we conclude that all the γ_i are zero, that is, $\varphi = 0$. The desired result now follows from Theorem 4.2.

Let u be a regular solution of (N^+). By Theorem 2.9 and (2.40),

$$u(x) = -\int_{\partial S} P(x,y)\chi(y)\,ds(y) + G(x), \quad x \in S^+, \quad (4.26)$$

$$G(x) = \int_{\partial S} P(x,y)\chi(y)\,ds(y), \quad x \in S^-, \quad (4.27)$$

where

$$G(x) = \int_{\partial S} D(x,y)\mathcal{Q}(y)\,ds(y), \quad x \in \mathbf{R}^2 \setminus \partial S, \quad (4.28)$$

$$\chi = u|_{\partial S}.$$

We re-arrange the elements of the subset $\{\vartheta^{(jk)}, j = 1, 2, 3, k = 1, 2, \ldots\}$ of (4.20) in the order

$$\vartheta^{(11)}, \vartheta^{(21)}, \vartheta^{(31)}, \ldots, \vartheta^{(1k)}, \vartheta^{(2k)}, \vartheta^{(3k)}, \ldots,$$

denote the new sequence by $\{\vartheta^{(m)}\}_{m=1}^{\infty}$, and use the Gram-Schmidt process to construct the orthonormal sequence $\{\eta^{(n)}\}_{n=1}^{\infty}$ in $L^2(\partial S)$. Thus,

$$\eta^{(n)} = \sum_{m=1}^{n} \kappa_{nm} \vartheta^{(m)}, \quad n = 1, 2, \ldots, \quad (4.29)$$

where κ_{nm} are well-determined numerical coefficients. Also, let $\{\tilde{f}^{(i)}\}_{i=1}^{3}$ be the orthonormalized set obtained from $\{f^{(i)}\}_{i=1}^{3}$.

We claim that $\{\tilde{f}^{(i)}, \eta^{(n)}, i = 1, 2, 3, n = 1, 2, \ldots\}$ is a fundamental orthonormal set in $L^2(\partial S)$. To convince ourselves of this we only need to verify that

$$\int_{\partial S} (\tilde{f}^{(i)})^{\mathrm{T}} \eta^{(n)} \, ds = 0, \quad i = 1, 2, 3, \ n = 1, 2, \ldots$$

But this is obviously true, since the $\tilde{f}^{(i)}$ and the $\eta^{(n)}$ are finite linear combinations of the $f^{(i)}$ and $\vartheta^{(jk)}$, respectively, and, by (2.71), (4.21) and Theorems 2.5 and 2.8,

$$\int_{\partial S} (f^{(\alpha)})^{\mathrm{T}} \vartheta^{(jk)} \, ds = \int_{\partial S} (T_{\alpha l} - x_\alpha T_{3l}) D_l^{(j)}(x, x^{(k)}) \, ds$$
$$= \int_{S^+} (A_{\alpha l} - x_\alpha A_{3l}) D_l^{(j)}(x, x(k)) \, da = 0,$$

$$\int_{\partial S} (f^{(3)})^{\mathrm{T}} \vartheta^{(jk)} \, ds = \int_{\partial S} T_{3l} D_l^{(j)}(x, x^{(k)})$$
$$= \int_{S^+} A_{3l} D_l^{(j)}(x, x^{(k)}) \, da = 0.$$

Without loss of generality, suppose that $n > 3$ and let

$$\chi^{(n)} = \sum_{i=1}^{3} \tilde{q}_i \tilde{f}^{(i)} + \sum_{r=1}^{n-3} q_r \eta^{(r)}, \qquad (4.30)$$

where

$$\tilde{q}_i = \int_{\partial S} (\tilde{f}^{(i)})^{\mathrm{T}} \chi \, ds, \quad i = 1, 2, 3,$$

$$q_r = \int_{\partial S} (\eta^{(r)})^{\mathrm{T}} \chi \, ds = \sum_{m=1}^{r} \kappa_{rm} \int_{\partial S} (\vartheta^{(m)})^{\mathrm{T}} \chi \, ds, \quad r = 1, 2, \ldots \quad (4.31)$$

Setting

$$u^{(n)}(x) = -\int_{\partial S} P(x,y)\chi^{(n)}(y)\,ds(y) + G(x), \quad x \in S^+, \qquad (4.32)$$

and using (4.26), just as in §4.1 we find that $u^{(n)} \to u$ as $n \to \infty$, uniformly on any closed subdomain $S' \subset S^+$.

From (4.27) it follows that

$$\int_{\partial S} P(x^{(k)}, y)\chi(y)\,ds(y) = G(x^{(k)}), \quad k = 1, 2, \ldots$$

By (2.25) and (4.21), this is the same as

$$\int_{\partial S} (\vartheta^{(jk)})^{\mathrm{T}} \chi\,ds = G_j(x^{(k)}), \quad j = 1, 2, 3, \ k = 1, 2, \ldots \qquad (4.33)$$

Applying Theorem 2.9 to $\tilde{f}^{(i)}$ in S^+, from (4.32) and (4.30) we now obtain the approximate solution in the form

$$u^{(n)}(x) = \sum_{i=1}^{3} \tilde{q}_i \tilde{f}^{(i)}(x) - \sum_{r=1}^{n-3} q_r \int_{\partial S} P(x,y)\eta^{(r)}(y)\,ds(y) + G(x), \quad x \in S^+,$$

where the first term on the right-hand side is a rigid displacement independent of n, $G(x)$ is given by (4.28), $\eta^{(r)}$ by (4.29), and the q_r are computed by means of (4.31), (4.33) and (4.28). Since the coefficients $\tilde{q}_i$ cannot be found in terms of the boundary data of the problem, we conclude that, in agreement with Theorem 2.43, the exact solution is determined in the limit up to an arbitrary rigid displacement.

4.3. The exterior Dirichlet problem

The construction of a fundamental sequence in the space of the solution for exterior problems meets with the usual difficulties that arise from the behaviour of the matrices $D(x, y)$ and $P(x, y)$ for $y \in \partial S$ and $|x|$ large. To overcome these obstacles, we need to establish some auxiliary results.

4.6. Theorem. *If S is a finite domain in $\mathbf{R}^2$, X a space of (3×1)-matrix functions defined on ∂S, Φ a linear functional on X, and*

$$F(x) = \Phi_y(D(x, y)\varphi(y)), \quad \varphi \in X, \ x \in S^-, \tag{4.34}$$

where the subscript y means that Φ operates with respect to y, then $F \in \mathcal{A}$ if and only if

$$\begin{aligned} p_\alpha &= \Phi_y(\varphi_\alpha(y) - y_\alpha \varphi_3(y)) = 0, \\ p_3 &= \Phi_y(\varphi_3(y)) = 0. \end{aligned} \tag{4.35}$$

Proof. Setting

$$D_{11}^\infty(x, y) = -a_2\mu^2(2\ln r + 2 + \cos 2\theta),$$

$$D_{22}^\infty(x, y) = -a_2\mu^2(2\ln r + 2 - \cos 2\theta),$$

$$D_{33}^\infty(x, y) = a_2\mu\big[\mu r^2 \ln r - 4h^2(\lambda + 2\mu)\ln r - 4h^2(\lambda + 3\mu)\big]$$

$$- a_2\mu\big\{y_1\big[\mu r(2\ln r + 1) - 4h^2(\lambda + 2\mu)r^{-1}\big]\cos\theta$$

$$+ y_2\big[\mu r(2\ln r + 1) - 4h^2(\lambda + 2\mu)r^{-1}\big]\sin\theta\big\},$$

$$D_{12}^\infty(x, y) = D_{21}^\infty(x, y) = -a_2\mu^2 \sin 2\theta, \tag{4.36}$$

$$D_{13}^\infty(x, y) = -a_2\mu^2\big[r(2\ln r + 1)\cos\theta$$

$$- y_1(2\ln r + 2 + \cos 2\theta) - y_2 \sin 2\theta\big],$$

$$D_{23}^\infty(x, y) = -a_2\mu^2\big[r(2\ln r + 1)\sin\theta$$

$$- y_2(2\ln r + 2 - \cos 2\theta) - y_1 \sin 2\theta\big],$$

$$D_{31}^\infty(x, y) = a_2\mu\big[\mu r(2\ln r + 1) - 4h^2(\lambda + 2\mu)r^{-1}\big]\cos\theta,$$

$$D_{32}^\infty(x, y) = a_2\mu\big[\mu r(2\ln r + 1) - 4h^2(\lambda + 2\mu)r^{-1}\big]\sin\theta,$$

where (r, θ) are the polar coordinates of x, for $|x|$ large we write (4.34) in the form

$$F(x) = \Phi_y\big(D^\infty(x, y)\varphi(y)\big) + \Phi_y\big((D(x, y) - D^\infty(x, y))\varphi(y)\big)$$
$$= F^\infty(x) + \tilde{F}(x).$$

Using (2.33), (2.34), (2.21), (2.23), (2.19), and (4.36), we find by direct calculation that $\tilde{F} \in \mathcal{A}$ and that

$$F_1^\infty = -a_2\mu^2\big[pr(2\ln r + 1)\cos\theta$$
$$+ p_1(2\ln r + 2 + \cos 2\theta) + p_2\sin 2\theta\big],$$

$$F_2^\infty = -a_2\mu^2\big[pr(2\ln r + 1)\sin\theta$$
$$+ p_2(2\ln r + 2 - \cos\theta) + p_1\sin 2\theta\big],$$

$$F_3^\infty = a_2\mu p\big[\mu r^2 \ln r - 4h^2(\lambda + 3\mu)\big]$$
$$+ a_2\mu(p_1\cos\theta + p_2\sin\theta)\big[\mu r(2\ln r + 1) - 4h^2(\lambda + 2\mu)r^{-1}\big],$$

which means that $F^\infty = 0$ when (4.35) hold.

4.7. Remark. Obviously, Theorem 2.17(ii) is a particular case of Theorem 4.6 for $\Phi\psi = \int_{\partial S} \psi\, ds$.

4.8. Theorem. *For any (fixed) $y \in \partial S$,*
  (i) $A(\partial_x)D^\infty(x, y) = 0$, $x \in S^-$;
  (ii) *the columns of $D - D^\infty$ belong to $\mathcal{A}$.*

Proof. (i) We can easily convince ourselves that the columns of D^∞ are generated by (3.28) and (3.36) with $l = m = 0$, $\psi = 0$, and Ω and ω

given respectively by

$$\Omega = -a_2\mu^2(\log z + 1), \quad \omega = -a_2\mu^2 z \log z,$$

$$\Omega = -ia_2\mu^2(\log z + 1), \quad \omega = ia_2\mu^2 z \log z,$$

$$\Omega = -a_2\mu^2[(z-\zeta)\log z - \zeta], \quad \omega = a_2\mu^2(\bar{\zeta} z \log z + 4h^2),$$

where $z = x_1 + ix_2$ and $\zeta = y_1 + iy_2$.

(ii) This assertion is proved by computing the entries of the matrix $D - D^\infty$ explicitly and verifying that its columns exhibit the far-field pattern (2.36) stipulated in the definition of the class $\mathcal{A}$.

Let the curve ∂S_* now be chosen so that it lies strictly inside the domain S^+.

4.9. Theorem. *The set (4.1), constructed as in Theorem 4.3, is linearly independent on ∂S and fundamental in $L^2(\partial S)$.*

Proof. Suppose that there are a positive integer N and real numbers c_i and c_{jk}, $i, j = 1, 2, 3$, $k = 1, 2, \ldots, N$, not all zero, such that (4.3) holds, and let ϖ again be defined by (4.4). Then $\operatorname{grad} \varpi_i = 0$ on ∂S, $i = 1, 2, 3$. Since $\varpi \in C^1(S_*^-)$, from the expression (2.11) we immediately see that

$$T(\partial_x)\varpi(x) = 0, \quad x \in \partial S. \tag{4.37}$$

Using the representation

$$\varpi = \varpi^\infty + \tilde{\varpi} + \sum_{i=1}^{3} c_i f^{(i)}, \tag{4.38}$$

where the functions ϖ^∞ and $\tilde{\varpi}$ are defined by means of the columns of the matrices D^∞ and $D - D^\infty$, respectively, from Theorem 4.8 and (4.37) we deduce that $\tilde{\varpi}$ is a regular solution of the exterior Neumann problem

$$A(\partial_x)\tilde{\varpi}(x) = 0, \quad x \in S^-,$$
$$T(\partial_x)\tilde{\varpi}(x) = -T(\partial_x)\varpi^\infty(x), \quad x \in \partial S,$$
$$\tilde{\varpi} \in \mathcal{A}.$$

According to Theorem 2.42(ii),

$$\int_{\partial S} (f^{(i)})^T T\varpi^\infty \, ds = 0, \quad i = 1, 2, 3.$$

Consider a circle Γ_R with the centre at the origin and radius R sufficiently large so that $\bar{S}^+ \subset \Gamma_R$ strictly. By Theorem 2.7 applied to $f^{(i)}$ and ϖ^∞ in $\Gamma_R \setminus \bar{S}^+$, the above equality yields

$$\int_{\partial \Gamma_R} (f^{(i)})^T T\varpi^\infty \, ds = 0, \quad i = 1, 2, 3.$$

Direct calculation now shows that these relations are equivalent to

$$\sum_{k=1}^N (c_{\alpha k} - x_\alpha^{(k)} c_{3k}) = 0,$$
$$\sum_{k=1}^N c_{3k} = 0.$$
(4.39)

Let Φ be the linear functional defined on the space X of bounded (3×1)-matrix functions on ∂S_* by

$$\Phi\varphi = \sum_{k=1}^N \varphi(x^{(k)}), \quad \varphi \in X,$$

and φ_c an element of X such that

$$\varphi_c(x^{(k)}) = (c_{1k}, c_{2k}, c_{3k})^T, \quad k = 1, 2, \ldots, N.$$

Then

$$\sum_{j=1}^{3}\sum_{k=1}^{N} c_{jk}\theta^{(jk)}(x) = \sum_{j=1}^{3}\sum_{k=1}^{N} c_{jk}D^{(j)}(x, x^{(k)})$$

$$= \sum_{k=1}^{N} D(x, x^{(k)})\varphi_c(x^{(k)})$$

$$= \Phi_y\big(D(x, y)\varphi_c(y)\big).$$

In view of the definition of Φ, (4.39) are equivalent to (4.35), therefore, by Theorem 4.6 and (4.38), $\varpi \in \mathcal{A}^*$. From (4.3) and (4.4) we then see that ϖ is the regular solution in S^- of the homogeneous Dirichlet problem

$$L(\partial_x)\varpi(x) = 0, \quad x \in S^-,$$
$$\varpi(x) = 0, \quad x \in \partial S,$$
$$\varpi \in \mathcal{A}^*.$$

By Theorem 2.15(i), $\varpi = 0$ in $\bar{S}^-$. Due to the analyticity of ϖ, we have $\varpi = 0$ in S_*^-, and the linear independence of the set (4.1) on ∂S is established by the argument used in the proof of Theorem 4.3.

Suppose now that the equalities (4.6) hold for some $\varphi \in L^2(\partial S)$. Since the points $x^{(k)}$ are densely distributed on ∂S_*, we deduce from Theorem 2.33(i,ii) that the elastic single layer potential V of density φ is a regular solution of the homogeneous interior Dirichlet problem

$$A(\partial_x)V(x) = 0, \quad x \in S_*^+,$$
$$V(x) = 0, \quad x \in \partial S_*.$$

Hence, by Theorem 2.15(i), $V = 0$ in $\bar{S}^+_*$. Due to the analyticity of V in $\mathbf{R}^2 \setminus \partial S$, we conclude that $V = 0$ in S^+, so that $TV = 0$ in S^+. Letting $S^+ \ni x' \to x \in \partial S$ along the support line of $\nu(x)$, we apply Theorem 2.36 to obtain the equation

$$\tfrac{1}{2}\varphi(x) + \int_{\partial S} T(\partial_x) D(x, y) \varphi(y) \, ds(y) = 0$$

for almost all $x \in \partial S$, the integral being understood as principal value. Theorem 2.47 now indicates that $\varphi \in C^{0,\alpha}(\partial S)$, with any $\alpha \in (0, 1)$. Hence, $V \in C(\mathbf{R}^2)$, which means that $V = 0$ on ∂S.

Now let Φ be the linear functional defined on the space $X = C(\partial S)$ by $\Phi\varphi = \int_{\partial S} \varphi \, ds$. Then the equality to zero of the integrals involving the $f^{(i)}$ in (4.6) is equivalent to (4.35), so, by Theorem 4.6, $V \in \mathcal{A}$. Since V is a regular solution of the homogeneous Dirichlet problem

$$A(\partial_x) V(x) = 0, \quad x \in S^-,$$
$$V(x) = 0, \quad x \in \partial S,$$
$$V \in \mathcal{A},$$

by Theorem 2.15(i), $V = 0$ in S^-. This implies that $(TV)^- = 0$, and, by Theorem 2.23, $\varphi = 0$. As in the proof of Theorem 4.3, we finally deduce that (4.1) is a fundamental set in $L^2(\partial S)$.

4.10. Remark. In classical three-dimensional elasticity [24] there is no need for the $f^{(i)}$ to be included in the set (4.1).

Let u be the (unique) regular solution of (D^-). By Theorem 2.44, we can write

$$u = \tilde{u} + \sum_{i=1}^{3} c_i f^{(i)}, \tag{4.40}$$

where $\tilde{u} \in \mathcal{A}$ and $c_i = \int_{\partial S} (g^{(i)})^{\mathrm{T}} \mathcal{R} \, ds$. By (2.41) and Theorem 2.11 applied to $\tilde{u}$,

$$\tilde{u}(x) = -\int_{\partial S} D(x,y)\psi(y) \, ds(y) + F(x), \quad x \in S^-,$$

$$F(x) = \int_{\partial S} D(x,y)\psi(y) \, ds(y), \quad x \in S^+,$$

where

$$F(x) = \int_{\partial S} P(x,y)\left[\mathcal{R}(y) - \sum_{i=1}^{3} c_i f^{(i)}(y)\right] ds(y), \quad x \in \mathbf{R}^2 \setminus \partial S, \quad (4.41)$$

$$\psi(y) = (T\tilde{u})(y), \quad y \in \partial S.$$

Since, by (4.40) and (4.41), $\tilde{u}$ is a regular solution of the exterior Neumann problem

$$A(\partial_x)\tilde{u}(x) = 0, \quad x \in S^-,$$

$$T(\partial_x)\tilde{u}(x) = \psi(x), \quad x \in \partial S,$$

$$\tilde{u} \in \mathcal{A},$$

it follows from Theorem 2.42(ii) that

$$\int_{\partial S} (f^{(i)})^{\mathrm{T}} \psi \, ds = 0, \quad i = 1, 2, 3,$$

which is equivalent to (2.77). This fact allows us now to proceed as in §4.1 and construct a similar scheme for the approximation of $\tilde{u}$.

4.4. The exterior Neumann problem

Let the curve ∂S_* and the points $x^{(k)}$ be as described in §4.3.

4.11. Theorem. *The set*

$$\{\vartheta^{(jk)}, \ j=1, 2, 3, \ k=1, 2, \ldots\}, \tag{4.42}$$

where the $\vartheta^{(jk)}$ are defined by (4.21), is linearly independent on ∂S and fundamental in $L^2(\partial S)$.

Proof. Suppose that there are a positive integer N and real numbers c_{jk}, $j=1, 2, 3$, $k=1, 2, \ldots, N$, not all zero, such that

$$\sum_{j=1}^{3}\sum_{k=1}^{N} c_{jk}\vartheta^{(jk)}(x) = 0, \quad x \in \partial S.$$

Representing ϖ defined by (4.23) in the form $\varpi = \varpi^\infty + \tilde{\varpi}$, where ϖ^∞ and $\tilde{\varpi}$ are constructed in terms of D^∞ and $D - D^\infty$, respectively, just as in the proof of Theorem 4.9 (this time with $\varpi \in \mathcal{A}$) we deduce that the set (4.42) is linearly independent on ∂S.

An argument similar to that used in the proof of Theorem 4.5 now shows that if

$$\int_{\partial S} (\vartheta^{(jk)})^T \varphi \, ds = 0, \quad j=1, 2, 3, \ k=1, 2, \ldots,$$

for some $\varphi \in L^2(\partial S)$, then the elastic double layer potential W of density φ satisfies $W = 0$ in S^+. Hence, as $S^+ \ni x' \to x \in \partial S$ along the support line of $\nu(x)$, Theorem 2.35 yields

$$-\tfrac{1}{2}\varphi(x) + \int_{\partial S} P(x, y)\varphi(y) \, ds(y) = 0$$

for almost all $x \in \partial S$, where the integral is understood in the sense of principal value. By Theorems 2.47 and 2.40, $\varphi(x) = 0$, $x \in \partial S$.

The fact that (4.42) is a fundamental set in $L^2(\partial S)$ now follows from Theorem 4.2.

The generalized Fourier series approximation $u^{(n)}$ of the (unique) regular solution u of (N$^-$) is constructed just as in §4.2, the procedure being simplified here by the absence of the rigid displacements $f^{(i)}$ from (4.42).

4.5. Numerical example

As an illustration of the generalized Fourier series method we consider the interior Dirichlet problem for an elastic disk where $h = 0.5$, $\lambda = \mu = 1$, ∂S is the unit circle with the centre at the origin, and the boundary conditions are

$$P_1 = 6x_1,$$
$$P_2 = -2x_2,$$
$$P_3 = 2(1 - x_1^2 + x_2^2), \quad x \in \partial S.$$

Let ∂S_* be the circle concentric with ∂S and of radius 1.5. We introduce polar coordinates with the pole at the origin and choose the points $x^{(k)}$, $k = 1, 2, \ldots$, on ∂S_* to be those corresponding to the polar angles

$$0, \pi, \tfrac{1}{2}\pi, \tfrac{3}{4}\pi, \tfrac{1}{8}\pi, \tfrac{3}{8}\pi, \tfrac{5}{8}\pi, \tfrac{7}{8}\pi, \ldots$$

Obviously, the set $\{x^{(k)}\}_{k=1}^{\infty}$ is densely distributed on ∂S_*.

Using Simpson's rule with 64 strips to evaluate the integrals over ∂S and following the computational procedure discussed in §4.1, we obtain the following approximate values for the solution at the points (0, 0), (0.05, 0), (0.5, 0.5), and (0, 0.95):

	(0, 0)	(0.05, 0)	(0.5, 0.5)	(0, 0.95)
$u^{(3)}$	0.000000 0.000000 3.666667	0.172004 0.000000 3.661667	1.618022 −0.951469 3.499989	0.000000 −1.770184 4.756414
$u^{(6)}$	0.001461 0.000000 2.918550	0.288200 0.000000 2.924077	2.803659 −0.736305 2.505363	−0.182989 −1.104677 3.578724
$u^{(15)}$	0.000000 0.000000 3.000188	0.303193 0.000000 2.992717	2.991931 −0.993187 2.501926	0.000000 −1.916326 3.789307
$u^{(27)}$	0.000000 0.000000 3.000000	0.300063 0.000000 2.992492	2.993269 −0.993391 2.499985	0.000000 −1.873686 3.791215
$u^{(51)}$	0.000000 0.000000 3.000000	0.300000 0.000000 2.992500	3.000139 −1.000260 2.499989	0.000000 −1.847026 3.788949
u	0.000000 0.000000 3.000000	0.300000 0.000000 2.992500	3.000000 −1.000000 2.500000	0.000000 −1.900000 3.902500

The last three rows of this array contain the exact values of the components of the solution u at the indicated points.

References

[1] M. Abramowitz and I. Stegun, *Handbook of mathematical functions*, Dover, New York, 1964.

[2] L. Bollé, Contribution au problème linéaire de flexion d'une plaque élastique, *Bull. Tech. Suisse Romande* **73** (1947), 281–285, 293–298.

[3] P.G. Ciarlet and P. Destuynder, A justification of the two-dimensional linear plate model, *J. Mécanique* **18** (1979), 315–344.

[4] D. Colton and R. Kress, *Integral equation methods in scattering theory*, Wiley, New York, 1983.

[5] C. Constanda, On the bending of micropolar plates, *Letters Appl. Engng Sci.* **2** (1974), 329–339.

[6] C. Constanda, Some comments on the integration of certain systems of partial differential equations in continuum mechanics, *J. Appl. Math. Phys.* **29** (1978), 835–839.

[7] C. Constanda, Sur les formules de Betti et de Somigliana dans la flexion des plaques élastiques, *C. R. Acad. Sci. Paris Sér. I* **300** (1985), 157–160.

[8] C. Constanda, Existence and uniqueness in the theory of bending of elastic plates, *Proc. Edinburgh Math. Soc.* **29** (1986), 47–56.

[9] C. Constanda, Bending of elastic plates, in *Integral methods in science and engineering* (Proc. Conf. Arlington, Texas, 1985), Hemisphere, New York, 1986, 340–348.

[10] C. Constanda, Fonctions de tension dans un problème de la théorie de l'élasticité, *C. R. Acad. Sci. Paris Sér. II* **303** (1986), 1405–1408.

[11] C. Constanda, Uniqueness in the theory of bending of elastic plates, *Int. J. Engng Sci.* **25** (1987), 455–462.

[12] C. Constanda, On complex potentials in elasticity theory, *Acta Mech.* **72** (1988), 161–171.

[13] C. Constanda, Asymptotic behaviour of the solution of bending of a thin infinite plate, J. Appl. Math. Phys. **39** (1988), 852–860.

[14] C. Constanda, Mathematical aspects of bending of plates with transverse shear deformation, Preprint no. 60, University of Bonn, 1988.

[15] C. Constanda, Differentiability of the solution of a system of singular integral equations in elasticity (*preprint*).

[16] C. Constanda, Potentials with integrable density in the solution of bending of thin plates, *Appl. Math. Lett.* **2** (1989), 221–223.

[17] C. Constanda, Smoothness of elastic potentials in the theory of bending of thin plates (*preprint*).

[18] C. Constanda, On Kupradze's method of approximate solution in linear elasticity (*preprint*).

[19] C. Constanda, Complete systems of functions for the exterior Dirichlet and Neumann problems in the bending of Mindlin-type plates (*preprint*).

[20] R.P. Gilbert and G.C. Hsiao, The two-dimensional linear orthotropic plate, *Appl. Anal.* **15** (1983), 147–169.

[21] A.E. Green and W. Zerna, *Theoretical elasticity*, Clarendon Press, Oxford, 1963.

[22] H. Hencky, Über die Berücksichtigung der Schubverzerrung in ebenen Platten, *Ing. Arch.* **16** (1947), 72–76.

[23] G. Kirchhoff, Über das Gleichgewicht und die Bewegung einer elastischen Scheibe, *J. Reine Angew. Math.* **40** (1850), 51–58.

[24] V.D. Kupradze et al., *Three-dimensional problems of the mathematical theory of elasticity and thermoelasticity*, North-Holland, Amsterdam, 1979.

[25] R.D. Mindlin, Influence of rotatory inertia and shear on flexural motions of isotropic elastic plates, *J. Appl. Mech.* **18** (1951), 31–38.

[26] C. Miranda, *Partial differential equations of elliptic type*, 2nd ed., Springer-Verlag, Berlin, 1970.

[27] N.I. Muskhelishvili, *Singular integral equations*, P. Noordhoff, Groningen, 1946.

[28] N.I. Muskhelishvili, *Some basic problems in the mathematical theory of elasticity*, 3rd ed., P. Noordhoff, Groningen, 1949.

[29] S. Prössdorf, *Some classes of singular equations*, North-Holland, Amsterdam, 1978.

[30] E. Reissner, On the theory of bending of elastic plates, *J. Math. Phys.* **23** (1944), 184–191.

[31] E. Reissner, The effect of transverse shear deformation on the bending of elastic plates, *J. Appl. Mech.* **12** (1945), A69–A77.

[32] E. Reissner, On bending of elastic plates, *Q. Appl. Math.* **5** (1947), 55–68.

[33] E. Reissner, On the theory of transverse bending of elastic plates, *Int. J. Solids Structures* **12** (1976), 545–554.

[34] E. Reissner, Reflections on the theory of elastic plates, *Appl. Mech. Rev.* **38** (1985), 1453–1464.

[35] P. Schiavone, Application of the boundary integral equation method to boundary value problems in the bending of thin micropolar plates (Ph.D. thesis), University of Strathclyde, 1988.

[36] P. Schiavone and C. Constanda, Existence theorems in the theory of bending of micropolar plates, *Int. J. Engng Sci.* **27** (1989), 463–468.

[37] P. Schiavone and C. Constanda, Uniqueness in the elastostatic problem of bending of micropolar plates (*preprint*).

[38] V.I. Smirnov, *A course of higher mathematics*, vol. 2, Pergamon Press, Oxford, 1964.

[39] Ya.S. Uflyand, The propagation of waves in the transverse vibrations of bars and plates, *Prikl. Mat. Mekh.* **12** (1948), 287–300.

[40] I.N. Vekua, *New methods for solving elliptic equations*, North-Holland, Amsterdam, 1967.

[41] A.J. Weir, *Lebesgue integration and measure*, Cambridge University Press, Cambridge, 1973.